Introduction to
PERMACULTURE

Introduction to
PERMACULTURE

Christopher Holland

Kruger Brentt
Publishers

2023

Kruger Brentt Publishers UK. LTD.
Company Number 9728962

Regd. Office: 68 St Margarets Road, Edgware, Middlesex HA8 9UU

© 2023 AUTHOR
ISBN: 9781787150447

For information on all our publications visit our website at http://krugerbrentt.com/

PREFACE

With growing awareness among people regarding the adaptation of sustainable farming practices and protecting the environment, several alternative ways of farming have become popular. One such practice which has become popular among people is permaculture. It is a climate-smart farming movement that would play a leading role in reducing the rapidly growing use of destructive, industrial-agricultural methods. It would be thus interesting to know what makes this practice climate-smart and sustainable in the long run. Permaculture is often confused with organic gardening. However, it is something more than that for it involves a comprehensive and dynamic system and can be practised at different levels and in various ways. Furthermore, it goes beyond farming practices and organic gardening because it integrates both the home and garden to create a system that has a lesser impact on the environment. Permaculture is an open-production system where energy is effectively utilized by one element and passed on to be used by another element before exiting the system whereas organic agriculture usually involves closed-production systems.

Permaculture aims to create stable, productive systems that provide for human needs, harmoniously integrating the land with people. The ecological processes of plants, animals, water, weather and nutrient cycles are integrated with human needs and technologies for food, energy, shelter and infrastructure. Unlike standard organic gardening, which is an improvement (but can still work against nature sometimes), permaculture can make deserts turn into oasis again, it thrives on creating new "food" forests rather than removing old ones and its goal is sustainability by design.

The present book contains eight chapters covering all related disciplines. These chapters include Permaculture; An Introduction, Natural Patterns & Permaculture Design, Houses, Water & Waste Management, Healthy Soil, Seed Saving & Nurseries, Home & Community Gardens, Farming, Forests and Tree Crops & Bamboo. This book provides a comprehensive introduction to permaculture. It elucidates techniques and their applications in the agriculture sector. It is written in a lucid style with theoretical discussions and being supplemented with illustrations, and tables for easy understanding of the subject. The book provides a step-by-step introduction to permaculture concepts, principles and design for sustainable agriculture. This book will be useful to all those who are interested in permaculture including, farmers, gardeners researchers, teachers as well as students and scholars.

We are grateful to all those persons as well as various books, manuals, periodicals, magazines, journals etc. that helped in the preparation of this book. In spite of the best efforts, it is possible that some errors may have occurred into the compilation and editing of the book. Further queries, constructive suggestions and criticisms for the improvement of the book are always welcome and shall be thankfully acknowledged.

Christopher Holland

CONTENTS

1

PERMACULTURE; AN INTRODUCTION

1.1 INTRODUCTION: AN OVERVIEW

- Permaculture is an ethical design approach rooted in observation of ecological processes. These processes act as a framework for creating regenerative systems for human material and non--material needs, including food, shelter, and energy, as well as economic, legal and social structures. Permaculture's hallmark is the beneficial integration of internal and external elements within a given space for optimal function, production and beauty. Systems designed using the permaculture approach mimic nature in order to minimize waste, maximize efficiencies, and produce abundant yields. Permaculture itself is not a discipline, but rather a design approach based on connecting different disciplines, strategies, and techniques.

- Since permaculture is still largely based on the research of its co--creators, Bill Mollison and David Holmgren, their writings are a primary source of introductory material on the topic and its practice. Today, permaculture is practiced worldwide and is based on Mollison and Holmgren's set of three permaculture ethics and twelve design principles. Permaculture ethics serve as the basis for decision-- making when developing a given system while design principles serve as the framework for implementation and management.

- Systems designed using permaculture can incorporate agricultural practices in addition to elements of a wide range of other disciplines, including landscape design, architecture, community development, energy production and storage, land management, and economic and legal structures.

- This ethical design approach challenges long held paradigms by holding up regenerative processes, production of a surplus, and care of people and the environment as objectives that are NOT mutually exclusive but instead, core tenets of good design. Permaculture can aid in our quest towards a viable, healthy, abundant future based on ecological processes and renewable resources.

- Permaculture is a design approach that utilizes multiple scientific disciplines and is practiced worldwide. Many organizations throughout the U.S. and around the world support and teach permaculture, including for--profit businesses, non-profits and higher education institutions. Today, permaculture organizations and programs exist in all 50 U.S. states including the Oregon State University Permaculture Design Program and the Permaculture Design Certificate offered at Cornell University. While close to half a century old, permaculture still has challenges to overcome before becoming fully accepted as a design approach in mainstream cultures. However, over the past decade it has flourished as individuals and communities have sought ways to fundamentally alter their relationship with the natural world and create fundamental systems--based change.

- The concepts and practice of permaculture offer one way to explore the broad scope of human impacts on the environment and how human behavior can be adapted to achieve greater harmony with these systems. Permaculture should not be viewed as a silver bullet, just as other forms of design, production and manufacturing are not viewed as cure--alls for the array of environmental and human health issues that exist and persist in the 21st century. Rather, permaculture is an approach focused on conscious, ecological design within the much larger environmental design toolbox.

1.1.1 Permaculture

The term permaculture stems from two words-permanent and agriculture. However, it incorporates a range of disciplines beyond agriculture, including landscape design, architecture, community development, energy production and storage, land management, and economic and legal systems. Through the use of multiple disciplines, permaculture considers and addresses a wide spectrum of issues. These topics are addressed as integrated parts of the whole and not as separate entities. This interconnectedness serves as permaculture's overarching ideology. Permaculture itself is not a discipline, but rather a "design approach based on connecting different disciplines, strategies, and techniques."

1.1.2 Issues Permaculture Considers

- Climate Change
- Food Security

- ⊙ Social Equity
- ⊙ Natural Capital Growth
- ⊙ Community Development
- ⊙ Ecological Processes
- ⊙ Energy Efficiency
- ⊙ Consumption
- ⊙ Waste Reduction
- ⊙ Renewable Resource Capture & Production
- ⊙ Human & Environmental Health
- ⊙ Biodiversity

Permaculture is an ethical design approach rooted in observation of natural systems that can act as a framework for creating productive landscapes for human needs. It utilizes whole--systems thinking to address human material and non--material needs including food, water, shelter, energy, and health (Figure 1). Permaculture design seeks to benefit all life forms (including humans, wildlife, plants, fungi and microorganisms). It seeks harmonious integration of human needs and ecological processes through systems designed to mimic natural environmental processes. Permaculture's hallmark is the integration of internal and external elements within a given space (i.e., geography, geology, soil biology, precipitation, sun patterns, climate, slope, social and cultural constructs, etc.) for optimal function, production and beauty.

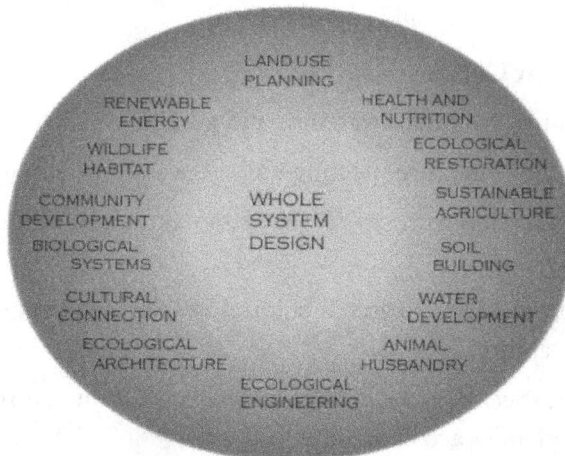

Fig. 1: Whole System Design, A Multidisciplinary Approach

Source: www.regenerativedesign.org/permaculture

According to permaculture co-creator Bill Mollison, permaculture is defined as, "the conscious design and maintenance of agriculturally productive ecosystems which have the diversity, stability, and resilience of natural ecosystems. It is the harmonious integration of landscape and people, providing their food, energy, shelter, and other material and non--material needs in a sustainable way." Through the implementation of conscious design, permaculture seeks to, "reduce the impact that human settlements have on nonrenewable and renewable resources, while creating an abundant living environment, catering to the needs of all living creatures".

1.2 HISTORY OF PERMACULTURE

- Bill Mollison was a biogeography professor at the University of Tasmania in the early 1970s when he, along with David Holmgren, one of his graduate research students, developed the concept of permaculture. Its roots lay in the rural landscapes of Australia where Mollison observed natural ecosystems and the interconnectedness of their ecological processes. These observations led in turn to the idea that productive landscapes designed by humans should mimic nature in order to minimize waste, maximize efficiencies, and produce abundant yields. These two individuals provided the origins of permaculture theory, and their writings are a primary source of introductory material on the topic and its practice.

- Permaculture practices were initially adopted by individuals' intent on achieving greater self--reliance and seeking more holistic landscaping approaches for their remote parcels of land. Since the 1970s, a larger world population, increased numbers of urban dwellers, and the growth of emerging markets have resulted in increased needs for energy resources and tangible goods. As a result, permaculture practices have become more widely adopted over the past 40 years in a wide range of urban and suburban settings.

Examples of Permanent Culture

- Personal Healing: personal awareness and growth
- Relationship Building: reconnecting individuals to one another
- Community Development: social, environmental and economic planning and development
- Regenerative Cultures: integrated cultural, social, environmental and economic growth and stability
- Restorative Justice: approach focused on mediation, personal needs, victim healing and offender accountability

Over time, permaculture has also evolved to become more inclusive of social and cultural issues. According to Holmgren, permaculture has transformed from a vision focused on permanent agriculture to one inclusive of permanent culture.

Thus, permaculture has evolved to embody both ecological processes and human cultures (Sidebar 2). Economic and social structures have been taken into consideration in order to build and repair human-- to--human, and human to natural environment, relationships. This transformation has expanded permaculture's reach socially and geographically – from individual to collective, and from rural to urban. The scope of 21st century permaculture focuses on the cultivation of permanent, thriving, ecologically-- minded communities.

1.3 REGENERATIVE VERSES SUSTAINABLE

- ◉ Currently, 40 plus years after initial development, permaculture is oftentimes referred to as "regenerative" design. It is important to understand what the word regenerative means and stands for within the permaculture context. A clear understanding of terms is especially important during a time in which the word sustainable has become a commonly used buzzword among environmentalists, politicians, academics, mainstream media outlets and private businesses, though not always for the same reasons or end goals.

- ◉ By looking at the root words, regenerate and sustain, the difference between the two can be seen. The Merriam--Webster Dictionary defines regenerate as, "restoring to a better, higher, or more worthy state." Thus, in the permaculture context, to regenerate is to ensure replenishment, rejuvenation, and re-- establishment of natural resources and social capital while ensuring the future security of natural resources, ecological processes and social structures. Sustain is defined as, "to bear up or keep in existence," or in the permaculture context, to maintain a systems' ability to provide resources at current production and consumption levels. To a certain extent "to sustain" is to create stability at current conditions and consumption levels while "to regenerate" is to produce resources for future resource security and stability using alternative methods of production.

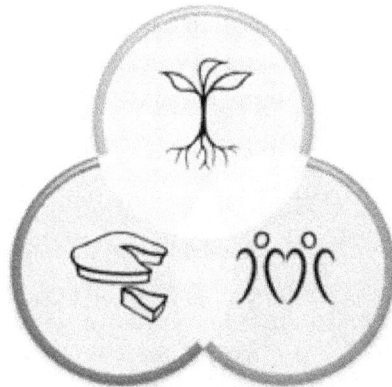

- ◉ Permaculture seeks to produce and provide an expansion of natural goods by using

Fig. 2: The Permaculture Ethics

abundant natural resources (air, water, sunlight, precipitation, healthy soil, etc.) as the system's main drivers of growth. Permaculture utilizes "waste" (or naturally generated excess) as an input to foster growth. Efficiency and self--sufficiency are important practices incorporated within permaculture designs in order to optimize production and consumption. In this way, permaculture is about regeneration and not dedicated to sustaining current lifestyle and consumption choices by simply applying alternative methods, materials and resources.

1.14 PERMANENT AGRICULTURE AND PERMANENT CULTURE

1.4.1 Permanent Agriculture

Is agriculture and animal management which improves the quality of land, provides income and produce, and is sustainable now and in the future.

1.4.2 Permanent Culture

Means conserving, supporting and working together with the local culture, while at the same time moving forward. Working with nature and people, as well as learning from them, and not working against or in competition with them.

- Permaculture helps us to understand and to create harmonic integrations between nature and people in the most sustainable way. Permaculture is appropriate for use in urban or rural locations, as well as for all scales of projects. Permaculture introduces traditional practices of nature management, integrated with appropriate modern technology. This is a holistic, kind, and environment friendly way for designing and building our natural living environment, as well as improving living standards, including housing, water supply, health, waste management, farming, energy, aquaculture, rivers, forests, livestock and much more.

- The term Permaculture was coined by Bill Mollison and David Holmgren in the 70s, and now is in practice in over 100 countries by thousands of Permaculture Design graduates.

- At this time there are many problems in the world, such as:

  - Damaged natural environments

  - Depleted and damaged farm land world wide

  - Polluted rivers, lakes, land, air and oceans

  - People, animals and plants are also becoming polluted, and many species are becoming extinct

  - Most of the world's population consists of very poor people, only a small percentage are very wealthy

- People have created all of these problems, and it is people who must change their ways for the earth to become healthy again. Action and change must come from all levels of society, including governments, businesses, workers, farmers, community groups, families, men, women, children, everyone! Future generations depend on this.

Fig. 3: People have created all of these problems

⊙ Permaculture offers techniques and ideas which help in directing us toward a healthier environment, cultures and people. This is based on certain ethics and principles. Permaculture ethics and principles provide a guide to being more responsible for our own lives, environment and future. As well as helping us to prepare a safe future for our families, culture, and natural environment.(Fig. 3)

1.5 ETHICS OF PERMACULTURE

Permaculture is still largely based on Mollison and Holmgren's three guiding ethics and twelve design principles. An ethic can be defined as "a set of moral principles, especially ones relating to or affirming a specified group, field or form of conduct." As an ethics-- based design approach, permaculture utilizes three guiding ethical principles or "permaculture ethics" that form the foundation for all design decisions. These ethics are largely based on practices and beliefs held in Australian aboriginal and other indigenous societies. By applying these ethics, permaculture seeks to respectfully utilize natural resources while benefiting all living species. The ethics can be seen as defining an inter-- related system – one in which no single one is superior to another in either importance or functionality, but wherein both the second and third ethical principles arise from the first.

The ethics of Permaculture are:

⊙ Care for the land

⊙ Care for the people

⊙ Care for the future

These ethics are explained as follows:

1.5.1 Care for the Land

Caring for the land means caring for our natural resources. Any action that damages, pollutes or destroys the environment or nature of Indonesia is also a loss for the people of Indonesia. Our natural environment must be protected and improved, this natural environment plays a key role in future of Indonesia.

Natural resources include:

⊙ Air

⊙ Flora: forests and plants

⊙ Fauna: wild animals, birds, etc

⊙ Water: lakes, rivers, springs, etc

⊙ Sea: beaches, coral reefs, marine life, etc

⊙ Land: farm land, including forests and land for animal grazing

If our land is managed in a sustainable way and slowly improved, productivity will also improve.

This will provide:

- ⊙ Long term productivity for farmers and their children
- ⊙ Protection and health for surrounding environments
- ⊙ Protection and health for those who farm the land

Fig. 4: Natural Resources

1.5.2 Care for the People

Caring for the people means preparing a healthy and safe future for everyone. Permaculture is about improving our opportunities, living environment, food supply, health and wellbeing (Fig. 5).

Sharing knowledge and assets will help us to:

- ⊙ Improve production, variety and quality of produce, as well food preservation and storage
- ⊙ Improve health and nutrition, including encouraging the use of effective natural medicines
- ⊙ Improve house health and hygiene, especially kitchens, air quality, toilets and waste management
- ⊙ Develop equal rights and opportunities for every individual; men, women and children
- ⊙ Improve livelihoods and work opportunities
- ⊙ Reduce daily hard work, such as carrying water,
- ⊙ Firewood, etc
- ⊙ Educate future generations in tradition, beliefs and knowledge, and in combining modern techniques with traditional culture

Fig. 5: Care of the People

1.5.3 Care for the Future

- What we do now affects the future. Caring for the future means always considering and planning for the future, not just 10 years, but 20, 50, 100 years in the future! For our grandchildren, and their children, are dependent on us to provide the best possible place for them to live.

- This ethic should be implemented by all levels of society, from governments and community groups, to families and individuals.

This can be implemented in ways such as:

- Protecting, distributing and marketing available resources

- Cooperation, not competition

- Protecting Indonesia's natural environment, by using renewable resources

- Reducing waste, by reusing and recycling

- Using less unsustainable materials

- Using renewable energy sources, such as solar power, hydroelectricity, biogas and wind power

- Managing population growth

Fig. 6 Care for the future

1.6 PRINCIPLES OF PERMACULTURE

- The principles of Permaculture should be implemented in every sustainable community design. These principles are an important guide for implementing Permaculture techniques. These principles also help to maximize efficiency and production in the most sustainable way, protect the soil, land, environment and people.

- Permaculture principles encourage creativity and maximise results. Every place is different, every situation and every family is also different. Therefore plans, techniques, plants, animals and building materials will be different each time. However, for every place and every activity, the same principles apply.

Diversity: Aims to integrate a variety of beneficial types of food, plants and animals into a design. This builds a stable interactive polyculture which provides for human needs, and other species needs as well.

Edge effect: In general, there is more energy and more diversity of life in the space where two systems overlap. The edge effect happens in this space because it receives benefits from both sides. Using the edge effect and other natural patterns creates the best effect.

Energy planning: Place elements within your design in a way that will conserve the most energy (this includes fertilizers, water and even human labour). Utilize the energy and resources that you have, first on site and later from the outside of the system, to save energy and money. Energy sources around us include natural energy forces, like gravity, wind power and water power (Fig. 7).

Energy cycling: In a natural system there is no waste or pollution. The output from one natural process becomes the resource for another process. Recycle and reuse resources as much as possible and as many times as possible.

Scale: Create human scale systems. Choose simple, appropriate technologies for use in designs. Create systems that are manageable, start small

Fig. 7 Energy sources around us

and take achievable steps towards and ideal goal.

Biological resources: Use natural methods and processes to achieve all tasks. Find materials in nature (plants, animals, bacteria) which support the system design and conserve the need for energy from outside the system.

Multiple elements: Support each vital and essential function in more than one way, so that if one element fails, it will not stop other elements in the process from functioning. Also, recognize that there is almost always more than one way to manage any process.

Multiple functions: Most things can be used in a variety of ways and for a variety of functions. One main rule in Permaculture is to try to design at least three uses for every element in a system. This will save space, time and money.

Natural succession: Work with nature and natural processes. Anticipate future developments through research and observation whenever necessary.

Relative location: Place every element of your design in relationships so that they can receive benefits from each other. For example, store tools near to the place where they will be used.

Personal responsibility: Our actions affect our own lives, our families' lives, our friends' lives and anyone else who has direct or indirect contact with us. Any constructive sustainable actions that we do will create benefits for many. The same is true of destructive actions, their affects will be felt far and wide (Fig. 8).

Fig. 8 Constructive sustainable actions

Cooperation not competition: Cooperation between people promotes community involvement, trading between members of the community, shared and improved knowledge and skills. Through cooperation many benefits can be achieved. Cooperation is important on all levels, in the family, in the village, in the districts and as a whole nation. Competition, on the other hand, creates conflict, jealousy and anger within communities especially if a resource is scarce. A good example is water use, usually the end result is that a few people have a lot while the rest receive only a little.

See solutions, not problems: Every problem that we are faced with has a solution. Often, the problem can contain within itself a solution. For example, turning weeds into compost and mulch, and using manure as a valuable resource for increasing soil fertility.

Observation: Natural patterns and cycles help us understand and make better plans for our farms, houses and gardens. Observation helps us to understand things like what works and what is not working and needs changing, by conducting simple experiments we can observe which are the best plants to grow and what is the best technique for growing them.

1.7 BEAUTY

Highly productive land can also be very beautiful, it is also the same for the house area. Indonesia has a very beautiful environment, and beautiful gardens and houses will add to it. Gardens and fishponds can be made in beautiful shapes. Flowers can be grown next to and among the vegetables. Small trees and legumes can be grown with fruit trees. This will encourage increased productivity and diversity.

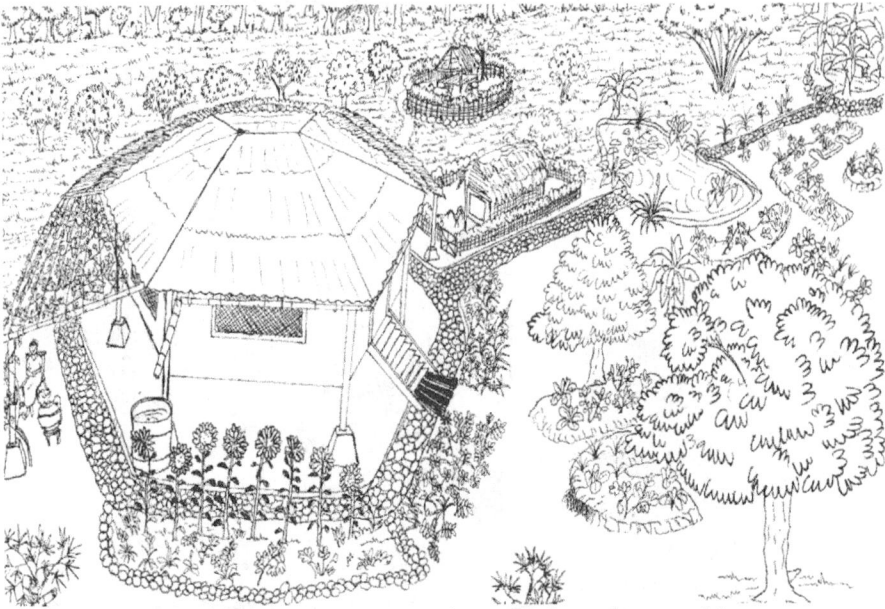

Fig. 9: Beautiful environment, beautiful gardens and houses.

1.8 PERMACULTURE APPLICATIONS

⊙ Permaculture incorporates knowledge from many fields of study in its designs and practices, and it can be utilized by and positively impact these fields. These areas of impact can be seen in Holmgren's Permaculture Flower depicted in Figure 4.

⊙ In this model, the permaculture ethics and principles are centrally located and are depicted as core to the practice, while petals depict areas in which permaculture designs and techniques may be implemented.

⊙ This particular diagram helps depict permaculture as a natural and social science that is about more than agricultural practices. Fields in which permaculture design can be utilized and incorporated include the building, technology, education, health and spiritual wellbeing, finance and economics, land tenure and community governance, and land and nature stewardship sectors.

⊙ In many instances, permaculture design may be implemented in these sectors through the use of cooperative models of teaching, ownership and governance.

⊙ Existing models include credit unions, grocery cooperatives and Waldorf school education (a non--sectarian and non--denominational pedagogical method based on self-governance that recognizes all world cultures and religions as equal and important).

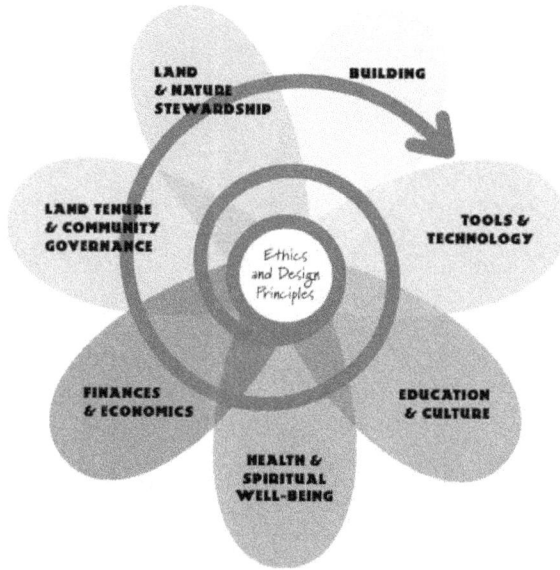

Figure 10. The Permaculture Flower

Source: www.permacultureprinciples.com/flower.php

2

NATURAL PATTERNS & PERMACULTURE DESIGN

2.1 PATTERNS

Planet earth is made up of patterns. Every aspect of the earth, from the smallest animal to the tallest mountain, contains patterns. Even the passing of time, in seasons and years, consists of patterns. Many patterns are repeated over and over again in different forms, some living and some not.

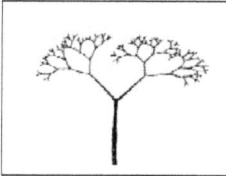

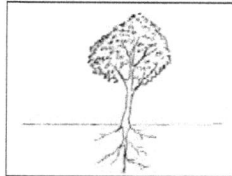

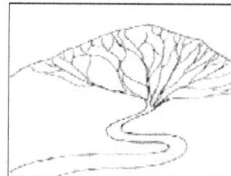

examples of natural flowing energy

Complex shapes are made up of simple patterns. Patterns are created in response to the natural flows of energy.

The patterns that exist in nature:

- ◉ Enable energy to flow
- ◉ Provide a solid structure
- ◉ Are natural responses to their surroundings
- ◉ Make life self sustaining and self perpetuating

We can either help energy to flow or we can stop it. People have also created many patterns, such as songs, music, dances, paintings, clothes, house designs and much more.

Traditionally, these patterns have been non-linear (not straight lines) and flow easily.

However, many human patterns, especially modern patterns, are not in harmony with nature. These patterns are often created in response to limited time and money, and create shapes that are unnatural and do not allow good energy flow.

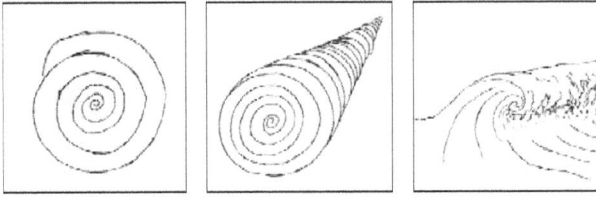

natural shapes

The result is that these patterns require constant maintenance and can cause problems and constraints. They also provide 'low quality' results, that lack beauty and do not feel good or comfortable, for example: box shaped houses, cities designed in squares, straight line agriculture, etc.

Are there any straight lines in nature?

Are there any straight or square lines in our bodies?

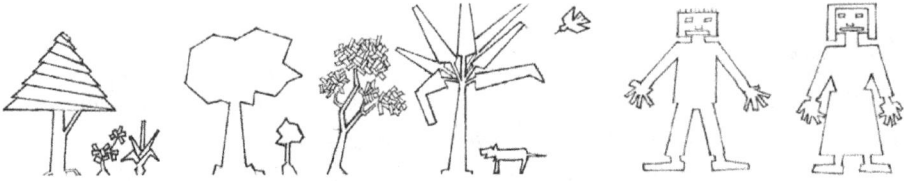

Which is stronger...

A curved wall or a straight wall?

A curved wall will support itself, while a straight wall needs support.

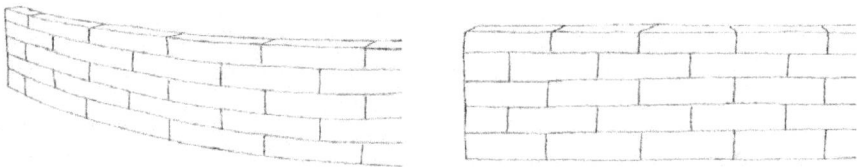

Every place has its own patterns, and so do the people who live there.

These patterns are unique! If we work with these patterns to create an environment or farming system, then we will achieve better results which require less maintenance.

Patterns in nature are very beautiful

Natural patterns in designs will also create beauty.

Natural shapes and patterns should be used as much as possible to improve beauty, especially around the home. In many cases this will improve productivity, while making an area much more comfortable to see and to work in, and give us a sense of pride.

Use your imagination!

Some of these patterns that are well known are the patterns on woven cloth. Other crafts also have beautiful patterns. Apply this knowledge of patterns to your garden.

Increasing Productivity

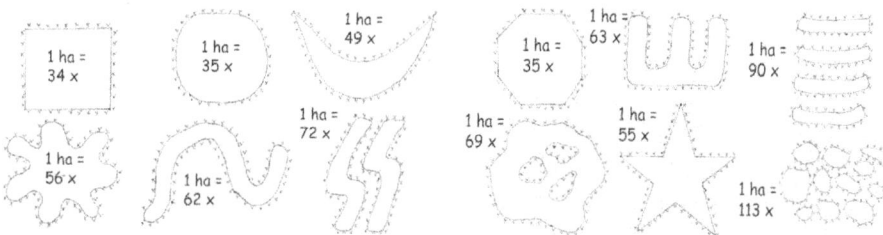

Changing the shape of a fish pond can affect and increase its productivity. Because the edge is the most productive area, if we increase the amount of edges, this will increase productivity.

Actually, the shape of the pond must fit appropriately with the shape of the land. This means that the land determines the shape, the shape does not determine the land.

If you work with natural shapes, you will achieve more productive results for less time and labour.

Because the pond contains more edge for the same amount of water, more trees, plants and water plants can be planted around this edge.

These trees and water plants can provide:

- ◉ Fish feed, in the form of fallen leaves and fruit
- ◉ Shade for the pond, which will reduce evaporation and help to regulate water temperatures
- ◉ Food for people
- ◉ Erosion control
- ◉ Material for making compost and mulch
- ◉ An increase of insects, birds and animals around the pond area, which will improve pollination rates and natural pest contr as well as provide more fish feed
- ◉ Healthier pond water

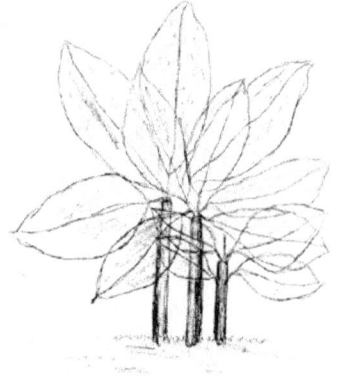

Fish ponds are a good example for working with natural patterns, but working with patterns and increasing edges will provide benefits for all types of agriculture and fish production.

Rice paddies created in alignment with natural shapes will be more efficient.

On slightly sloped land, all kinds of patterns can be applied to make use of rain water during the wet season.

Vegetables and soil can be protected, while water run-off can be reused.

Edges occur naturally on the land, but we can also create them. All edges can be used, and all usage can increase production and diversity. All paths have edges on both sides, which are often not used for production. Even just planting these path edges with fruit and flowers will provide many benefits. These benefits could be extra income, mulch, and more birds and insects. Because what you plant is along a path, it will be easy to harvest the produce!

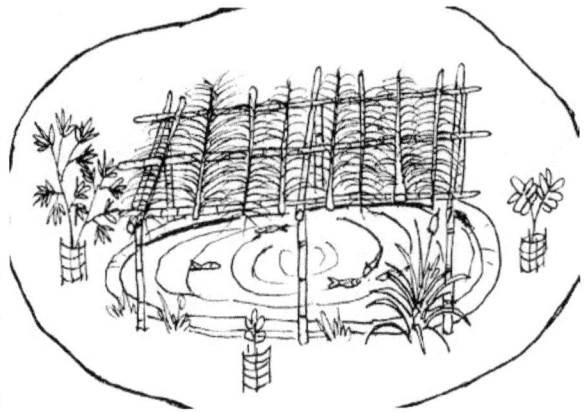

2.2 METHODS OF DESIGN

It will cover some ideas and methods used in planning long term designs for agriculture land, animal management, houses and other projects.

"Where will the element go?"

"How can it be placed to provide maximum benefits for the system?"

In designing any system, Permaculture combines a series of techniques and strategies. Techniques are how you do something. Strategies are how and when you do these things. Design is about making a pattern from different elements and working with the land to create a system.

Creating a long term design is very important because:

- It helps to plan for the future
- It helps you to see what are priorities for developing a successful farm or project
- It enables you to see how to integrate different parts of a system to save resources and labour, while increasing productivity
- It allows you to plan how to use waste from one section as a resource for another section

- ⦿ It reduces long term labour needs and helps you achieve maximum benefits from your work

- ⦿ It speeds up the development of a farm or project, by using the appropriate technique at the right time

- ⦿ You can make plans for extreme weather conditions, like a storm, drought or flood. This is very important! There are many techniques in this guidebook to help you prepare for these extreme conditions

A plan provides you with a frame. Like a house, the framework is necessary before you begin to build the rest of the house, and good framework will create a strong and long lasting house.

A plan will bring more order, but can still have flexibility. Plans can change as circumstances change. If you have successes or failures, adjust your plans accordingly.

2.3 APPROACHES TO DESIGN

2.3.1 Maps

"Where is everything?"

"What is the shape of the land?"

Drawing or making a map of the land is a good way to see where everything is in one picture. An overview map is drawn as if you are looking down at the land from above, like an eagle looking down from the sky.

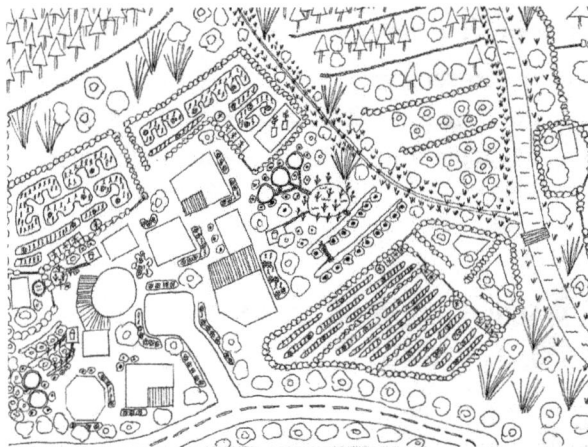

KEY

H
K

◎ Fruit

🖌 Vegetable

▭ Flower

⊡ Washing

▥ Shade

Ⓢ

∞∞∞∞∞ Living

🌀 Waste water

⬭ Animal
pens D-duck

∞▭∞ Fence

⊞
❋❋❋

❋ Liquid

🌲🌲🌲 Tree
☁

ℓ ℓ ℓ
❀ ❀

🌱

∞∞∞∞∞∞

🦆 Duck

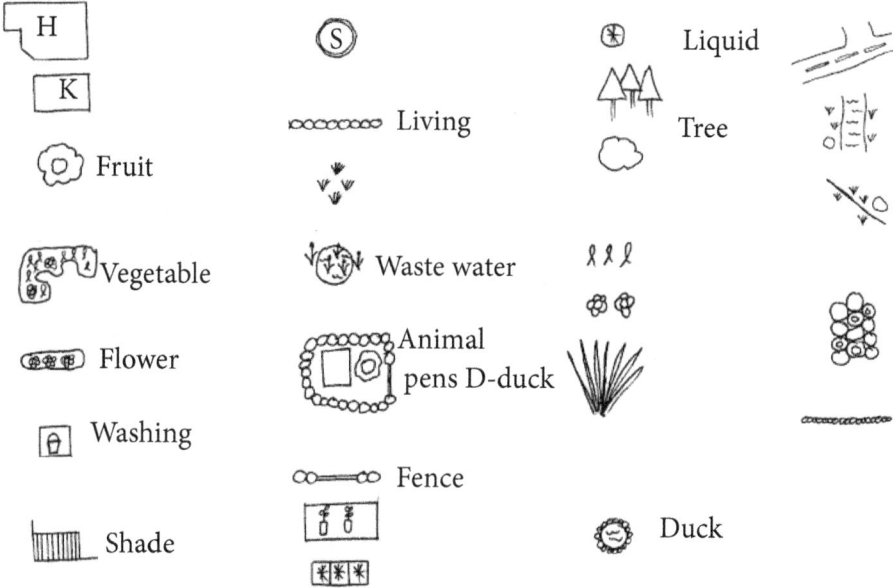

A map can also be made with sand or soil, using models to represent relevant features of the land. This method is often easy and fun. You can repre sent mountain slopes and rivers, and even experiment using real flowing water.

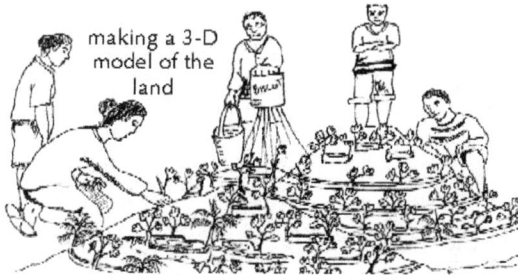

making a 3-D
model of the
land

A profile map (cross section map) is a different way of looking at the land, this is very useful for looking at land with slopes or at a specific section of the land. This is like cutting a slice of cake and then looking at the slice from a side view. Land surface, including buildings and trees make up the top of the slice, with tree roots growing into the slice of cake.

A water catchment profile

All different parts of a map should be drawn "to scale". This means that they are the same size in relation to each other as they are in real life.

To achieve this:

- Measure every section with equal sized steps.

- Count the number of steps for each measurement.

- Compare distances of different sections. A wall that measures 20 steps should be drawn twice as long as one that measures 10 steps. A garden plot that measures 25 steps should be drawn 5 times as long as one that measures 5 steps.

- Draw the shape of the area and write the actual measurement beside it.

These drawings don't have to be perfect, but using this method will help you to draw a more accurate map. A more accurate map will help to make better designs.

Show where land is flat, gently sloped or steeply sloped. Different techniques and strategies will be used for different sections of the land, so it is important to show their differences. Also take note of river gullies, caves or any other uncommonly shaped land formations.

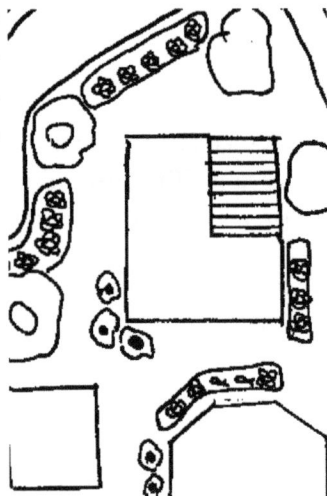

Maps should include:

- Existing buildings
- Future building plans
- Existing vegetation
- Future garden plans
- Existing animals and animal shelters
- Future animals and animal shelter plans
- Rivers, creeks, ponds and water flows
- Roads and paths
- Flat land, gently sloped land and steeply sloped land
- Pipes for water and electricity
- Boundaries and fences
- Any cursed land or sites
- Land that is subject to extreme problems, such as erosion, flood planes, rocky ground
- And most importantly, a 'key'

All the different features on the map are given [...] map 'key' is a section of the map where all the lette[...] The map 'key' acts like a key in real life; it unlock[...] presented on the map.

To make the map easier to read, use different colors for the different features. For example, use yellow for water, green for buildings, red for roads, etc.

Different colors may also be useful to distinguish existing features and future plans. For example, use black for existing features and red for future plans.

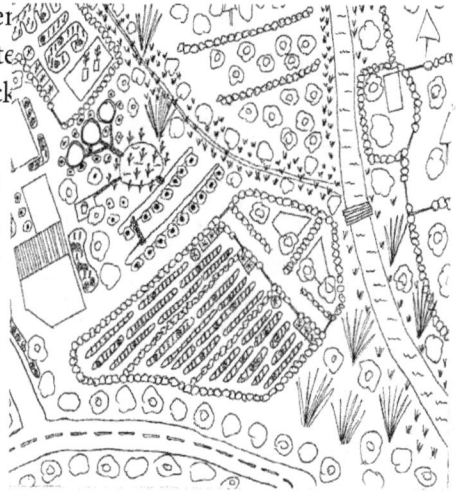

2.3.2 Element Analysis

How does Everything Work Together?

A simple "Needs and Products" table is a very important and a very easy way of understanding:

"What do we need for each element?"

"What products does each element give us?"

For example, if you keep chickens in a chicken yard, you will achieve many benefits, besides just meat. To make a chicken yard and have healthy chickens, we have to know what the chickens need. Only then can we consider what products they can provide.

CHICKEN NEEDS feed, shelter, water, protection from predators, shade, medicines, friends (other chickens), dry earth, fencing, box for laying eggs, fresh air

CHICKEN NEEDS
feed, shelter, water, protection from predators, shade, medicines, friends (other chickens), dry earth, fencing, box for laying eggs, fresh air

CHICKEN PRODUCTS
meat, eggs, manure, money, feathers, work (weed and pest management)

Another example is the needs and products of a vegetable garden.

VEGETABLE GARDEN NEEDS
seeds, compost, liquid fertilizer, mulch, healthy soil, fencing, nursery, cultivation (tools and labour), water, sunlight, weed and wind management

VEGETABLE GARDEN PRODUCTS
vegetables, fruits, herbs and spices, compost material, animal feed, flowers, money, other products to trade, mulch material, windbreak

We can use the "needs and products" table to connect different elements within a system and reduce costs and other outputs.

For example, chicken feed can come from:

- ◉ Kitchen food waste (garden products via the house)

- ◉ Weeds (garden product)

- ◉ Pruning from trees (garden product)

- ◉ Rotten food (garden products via house)

- ◉ Diseased plants (garden products, giving them to chickens will stop disease from spreading)

- ◉ Insects and bugs (product from building the chicken yard)

Products from the chickens can then fulfill the needs of other systems, for example:

- ◉ Eggs, meat, money (needs of people)

- ◉ Feathers (needs for cultural ceremonies, handicraft material, bedding material)

- ◉ Chicken manure (needed for making compost to be used in the garden)

- ◉ Work (needed for managing weeds and as a 'chicken tractor')

Often many different needs can be fulfilled by the same source, for example trees around a chicken yard can produce:

- Food for people
- Chicken feed
- Shade for chicken and people
- Windbreaks
- Medicines
- Fence posts
- Mulch

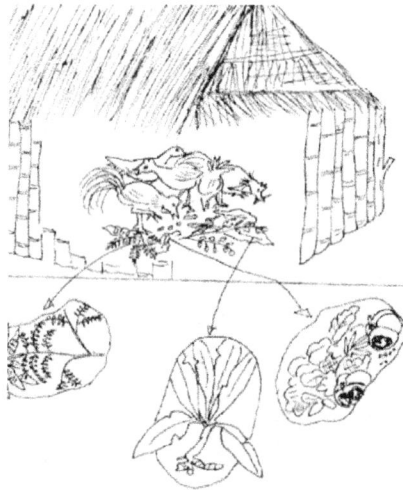

In this way we can make a needs and products analysis for anything.

Another example is buffalo:

BUFFALO NEEDS
feed, water, shade / shelter, grazing area, medicines, plough and training (for working the land), rope, waterhole

BUFFALO PRODUCTS
meat, money, manure, weed management, horns, ploughing and fertilizing the field, leather, bones, transportation (for farm produce), cultural ceremonies

buffalo ploughing the contour on sloped land

2.3.3 Zones

Creating a zone map can help to reduce the amount of time and labour needed to create a Permaculture system. This map helps to show how to combine compatible elements with what they need, in the most efficient way.

To make a zone map, you can look at the farm as five zones, the zones start at the house. Each zone represents a different area of the system. Different elements are placed in each zone, depending on:

- ⊙ How often that element is visited - Zone 1 is the most visited, while zone 5 is the least visited

- ⊙ How much maintenance is required - Zone 1 requires the most maintenance, while zone 5 requires the least

- ⊙ Access to water supply - Zone 1 requires the most access to water, while zone 5 requires the least

- ⊙ The amount of land area required - Zone 1 requires the smallest amount of land, while zone 5 requires the largest amount

- ⊙ Compatibility with the land - Their compatibility with other elements in their surroundings

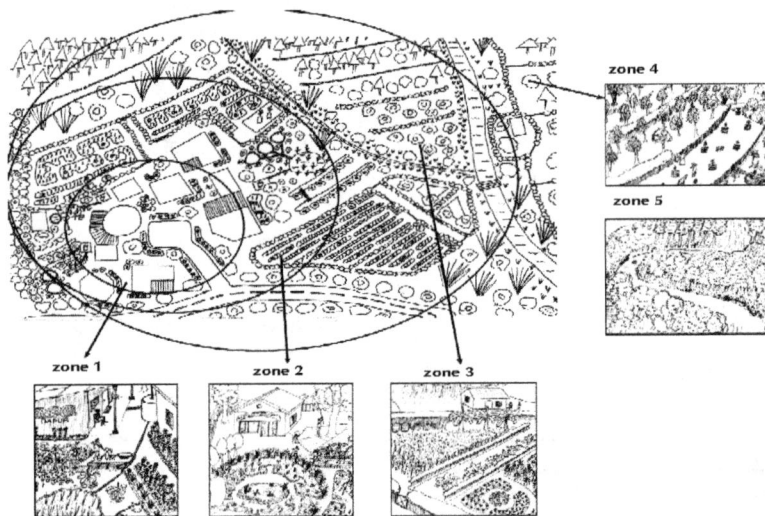

Zone 1: The Home Garden

Zone 1 is the area closest to your house. Within zone 1 place whatever needs the most attention and maintenance, and will also provide daily household needs. Elements within zone 1 could be:

- The house, kitchen, washing area, toilet
- Nursery
- Pergolas and vines
- Home and medicinal gardens
- Fruit trees which provide shade
- Water pump
- Compost heap
- Water tank
- Waste water pond
- Aquaculture ponds
- Smaller livestock

Once you have all the elements required for zone 1, the important task is to place them together in the most appropriate and compatible way.

Plants which provide shade are good for near the house, but too much shade on a vegetable plot could reduce production.

The water pump should be placed far from the toilet area, to avoid water contamination from toilet wastes.

Pergolas for planting vines will provide a shaded, cool area around the outside of the house. In very hot environments, place the pergolas on the west side of the house, and for colder environments, place the pergolas on the eastern side. This will keep the house cooler by reducing direct sunlight on the house.

Elements that share needs or use what another element provides, should be placed together. This will maximize benefits, while minimizing time and labour. This concept should be applied for all zones, and to how one zone relates to another.

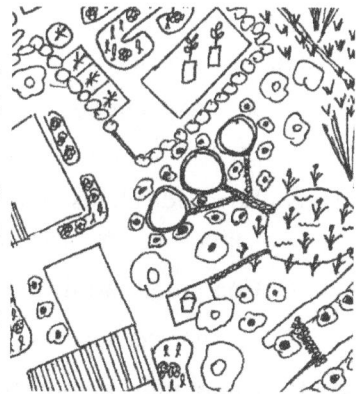

Zone 2: The Village

The village includes all infrastructure, such as roads, paths, religious buildings, schools, village land, government offices, and community housing.

The entire village can be built well using Permaculture designs, which can include the following elements:

- ⊙ A village nursery and garden, which can be placed on the elementary school grounds

- ⊙ Road side plants, to provide shade, animal feed, protection from wind and dust, fruits, medicines, and firewood

- ⊙ Community waste management facilities

- ⊙ Clean water sources

- ⊙ Community demonstration plots and a food supply gardens

- ⊙ Markets, designed in a way to avoid disorganization and bad odors, with minimal maintenance needs

Livestock management, by tying or providing shelters to avoid the animals damaging home gardens and community food production systems. A place to keep animals such as goats or buffalo during the night. This place will also make it easier to collect the animal manure

Elements in zone 2 can also attract and interest tourists, for example:

- ⊙ Stands selling produce from home gardens (kaki lima), to attract tourists

- ⊙ Overnight accommodation for tourists, which can also bring in extra income

- ⊙ Promotion of local tourism attractions

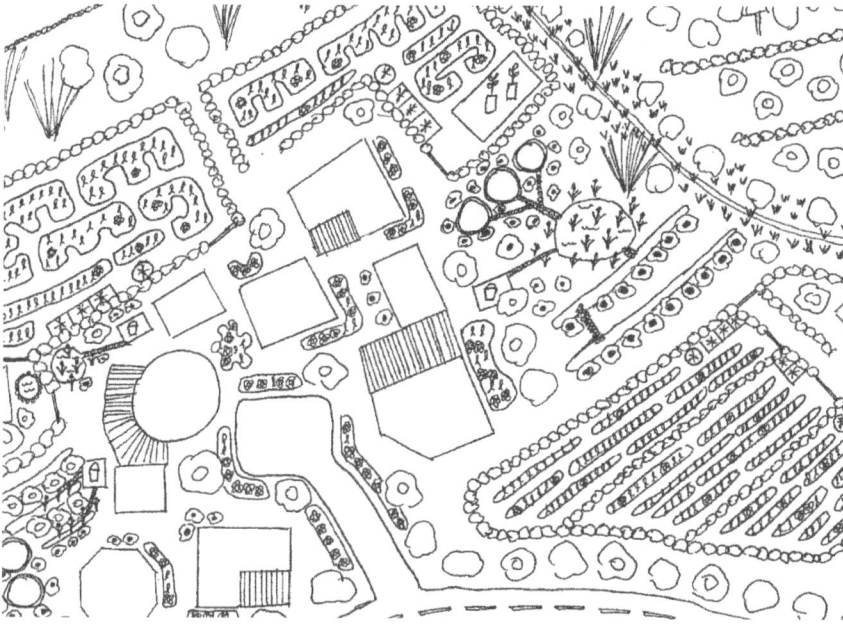

Zone 3: Small Farms at the Edge of the Village

Zone 3 is land which requires less attention and maintenance; therefore, it is usually located further away from the house.

Zone 3 elements include:

- Land which has permanent fencing
- Polyculture (integrating different crops together)
- Tree crops and annual crops, for year-round food security
- Land divided into sections for crop rotation and animal grazing, with fodder trees as the land borders
- Swales and terraces to protect the land from erosion
- Emergency crops planted in case of a famine or other disasters
- Use non-hybrid seeds
- Do not use chemicals, and do rotate crops to avoid stripping the soil of nutrients
- Reuse all organic wastes for compost and stop burning land
- Integrated Pest Management (IPM), use companion planting to reduce pest problems
- Clever designs and strategies will work better than using expensive farm machinery

Zone 4: Community Forests

This zone is located around the village farms.

Here community resources are grown and burning is not permitted. A variety of crops can be grown in this zone, from controlled orchards to semi-wild forests.

To avoid conflict, zone 4 should be designed and managed by the community. It requires minimal maintenance and attention, meaning minimal watering and fertilizing.

This zone requires more land area and it includes tree crops, which if planted in zone 1, 2 or 3, would decrease the productivity of other crops because of too much shade and root competition.

Zone 4 elements include:

- Fruit, oil and nut trees
- Firewood and timber trees
- Bamboo
- Buffalo, cows and goats
- A water source
- Coffee plantations
- Medicinal plants
- Paddies, if there is enough water
- Swales for reforestation
- Plants for craft materials

Zone 5: Conservation Forests

Zone 5 is land which is left untouched. Natural forests provide many functions, such as food and medicines for people, food and habitats for native animals, and protection against hunting and erosion. Conservation forests can also produce money for the community while being kept preserved for future generations. Logging forests will only provide short term benefits for a few people, while causing long term damage for many people.

Zone 5 is usually located furthest from residential areas and can be on land which is more difficult to use for cultivation, such as steep slopes, eroded waterways, mountains and rocky ground. Plants which are grown on this land should be native plants, with a variety of sizes, ages and species.

Zone 5 elements include:

◉ Forest conservation laws which are agreed upon and enforced together

◉ Ecotourism projects

◉ Non-timber forest products

- ⊙ Low-impact activities are allowed, but burning is not permitted
- ⊙ Forest rangers can be appointed to protect the forest

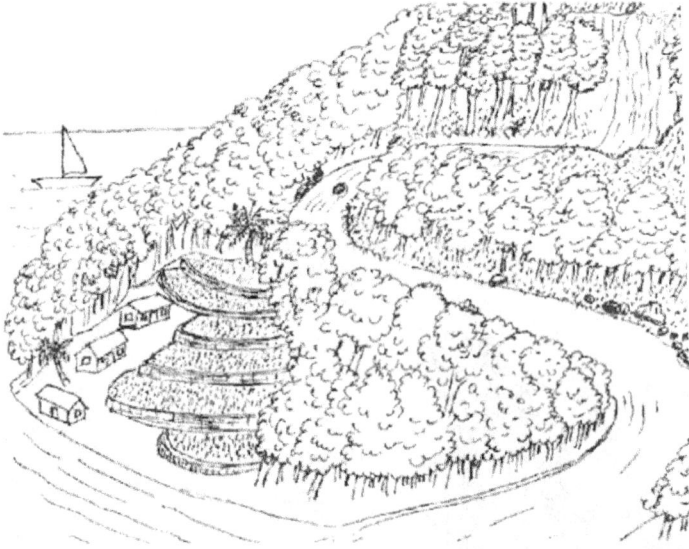

There are some different situations which could affect how land is zoned, such as:

Access to water. If there is access to a spring or irrigated water the land will be much easier to use for intensive crops, fruit trees or animals

Access to roads. If the land has access to roads, produce can be easily transported. Therefore, large amounts of crop produce will not be wasted

Erosion levels. If erosion already exists or the land is very steep, the soil must first be stabilized before intensive agriculture can be attempted, this can be done using terraces, swales or by planting trees

Soil quality. Very poor soils or very rocky soils must be worked on for many years to become fertile, or tree crops should be the main production focus. Usually, it will be better and easier to start by planting tree crops, and then eventually begin growing small plots of other crops and vegetables. Pruning back the tree crops will also help to improve the cropland faster

SMART IDEAS!

- ⊙ Some elements could be located in more than one zone, like corn, citrus and pigs. This will depend on:
  - ▪ The type, quality and size of the land
  - ▪ The techniques and strategies used
  - ▪ Which crops are for selling and which are for the family to eat
  - ▪ The possibility of integration with other elements

⊙ Making pathways is very important because they can connect zones in efficient ways. The pathways will provide location points for liquid compost, animal pens and water access. All pathways can border or fenced with production crops. These path borders can be small garden plots, flowers, herbs, vines or smaller fruit trees. This will increase the use of non-productive land and make the farm more beautiful!

The zones can also be implemented on a community or village level. This idea can save time, costs and labour. If different farms are working together, the production process will be more efficient, resources can be continuously reused and everyone will benefit

Exercise: Draw a zone map with only the basic zone outlines. On a separate piece of paper, draw and cut out the different elements (like houses, vegetable plots, a chicken yard, ponds, etc). Place these elements on the map in any way you like to design your own farm. Explain how the different placed elements are connected to each other.

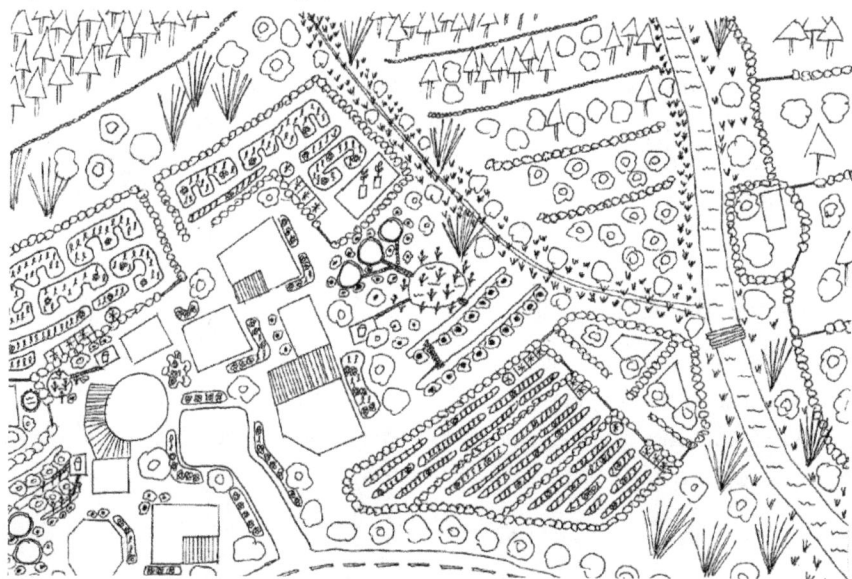

2.3.4 Sectors

Sectors look at the natural factors that affect the land and the production levels of the land. These natural factors are sun, wind, water flow and flooding potential, fire, slopes, soil types and sacred lands. Sector planning is done to channel these natural factors into or out of a system.

The knowledge gained from understanding the effects of these natural factors leads to planning that will:

⊙ Help to maximise yields

⊙ Reduce mistakes made, for example, planting crops or trees which will get washed away with the next heavy rains or floods

⊙ Make the farm more resilient and capable of facing disasters and extreme conditions, like fires, flooding or erosion

2.4 SUN

The direction of the sun is important. By observing its path during the day you will find where the maximum and minimum sun exposure areas are. Remember that this changes from wet season (higher arc in the sky) to dry season (lower arc in the sky).

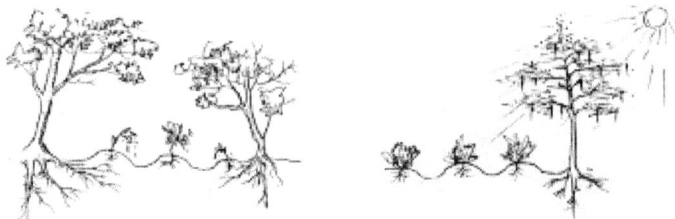

Too much shade   **Enough Sunlight**

Use areas of maximum sunlight exposure first and to plant the most important crops. For reforestation it is also important to establish the sunniest areas first. The more shady areas are better suited for keeping animals. But some crops, like coffee and vanilla, will actually grow better with some shade.

2.5 WIND

⊙ Where does the wind usually come from and how strong is it?

⊙ Plant windbreaks in the appropriate areas to protect crops, animals, aquaculture and the house area. In very exposed areas, only plant tough, strong trees, because the wind and sun can dry out and damage many trees. Strong winds can also reduce crop growth and increase water usage.

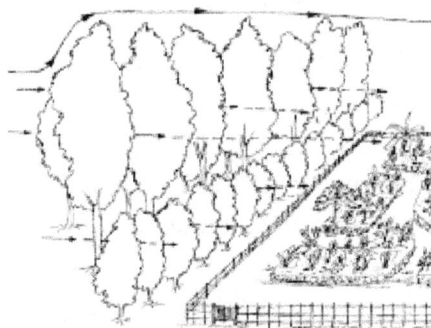

Windbreak tree crops

2.6 WATER FLOW

⊙ Where does water flow through the land? Are there any springs? Are there areas of land which experience erosion?

⦿ It is important to protect natural water courses and springs by planting vegetation or trees which will also prevent erosion. Water collection points and irrigation can be established to channel water.

⦿ Erosion can be prevented by using swales and planting trees. This will also prevent potential landslides and large scale erosion, which if unchecked could become a huge problem. Remember that every time erosion

Repairing eroded waterways

happens, valuable topsoil is lost and the chances of mudslides increase. Erosion also causes problems for rivers and oceans.

2.7 FLOODING

⦿ Are there areas of the land which flood during hard rains? Are there areas which are naturally swampy or where water overflows?

⦿ Observe where water comes from and protect these areas from erosion and landslides. The best way to reduce flooding and water runoff is to use swales, terraces and reforestation to store as much water as possible in the ground.

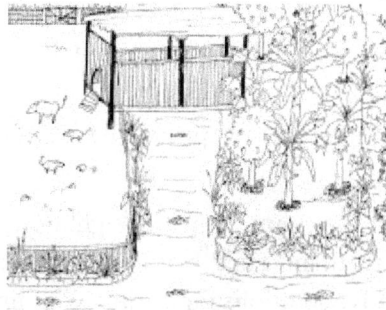

Take advantage of naturally swampy or flood prone areas by planting crops which are compatible, such as rice, kangkung and taro. Ducks, fish and fresh water prawns can also be produced in these areas. In this way, water will be stored and used, and excess water can be regulated.

2.8 SLOPES

⦿ How steep are the slopes on the land? How can the soil be protected and how much of the slope can be used for agriculture production?

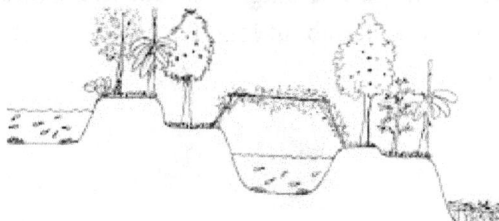

- As with flooding, catching and storing water in the soil will improve sloped land agriculture and provide protection for the soil. Different techniques such as swales and terracing can be used. Gravity can also be used for irrigation, this can be done naturally by using swales, or by using piping, bamboo and hoses. Gravity can also be used to run water into aquaculture systems or water catchment systems.

- From what direction does fire usually come from? Usually, fire will move most quickly up slopes and from the direction the wind most often comes from.

- To help reduce or stop fire from spreading, plant a firebreak, this could be two or three rows of fire resistant plants with cleared gaps on either side (like a wall or fence). Plants which are fire resistant include banana, papaya, fig, cactus, etc.

Protecting gardens from fire

- These plants should be grown near areas where the fire might come from. A firebreak can also be multifunctional, it can provide food, wood and other resources.

- Firebreaks are very important for protecting buildings, animal pens, vegetable plots and other intensive agriculture areas.

- It is also important to communicate with neighbours about the danger of fire and to work together with them. Hold community meetings about how to prevent fires and find alternative solutions for dealing with fire.

2.9 SOIL TYPES

- Are there different types of soil on the land? Are there differences in the depth of soil hummus?

- Areas that are rocky, swampy or salty should be given extra attention and be treated differently. Test the soil to find out what types of soil you have. All types of soil can be improved and changed into more productive and healthy soil if good management is used. Use tough trees for rocky or salty soils, water plants for swampy areas, and think about long term ideas which can make these areas more productive.

2.10 SACRED OR CURSED LAND

- ◉ Are there any sacred or cursed areas on your land which may affect what is done to the land?

- ◉ Discuss these matters with the local community leaders and spiritual leaders to find the best solution for using this land, to heal the land (possibly through a ceremony), or to find out if it is better for the land to be left uncultivated.

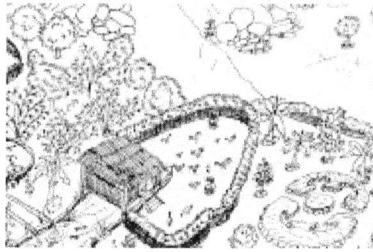

2.11 OBSERVATION AND DATA COLLECTION

Observation is very important and should be the first step taken when planning agriculture projects for your land. Through observation we can see how natural elements affect the land. For example, the same variety of tree will grow differently in one area compared to in another area. Is this because of the amount of water available, different soil depths, wind exposure, sunlight exposure or another factor entirely? Observation can show us and help us to make better plans.

Nature gives signs that we can look for, such as:

- ◉ Plants with fleshy or fuzzy leaves will grow better where there is more water available

- ◉ If there are often strong winds, all tall trees will grow leaning to the opposite direction, and plants will grow smaller and stunted.

- ◉ Yellowing of leaves and new growth, early maturing and smaller fruit and flowers are all signs of nitrogen deficiencies in the soil. If you observe and work with nature, you will save time, labour and expense.

2.12 LOCAL KNOWLEDGE

Local knowledge is always an important source of information. Much traditional knowledge is passed orally and not written down. Collect as much information as possible about climate, natural factors, what grows well and what used to grow well, to help to reduce mistakes. The elders in a community are the best sources of information. This kind of information can be very important when planning for extreme weather conditions.

2.13 THE LOCAL GOVERNMENT

- ⊙ Government agriculture workers can help to provide some information and support. Information about government projects, weather patterns, seeds and plants, techniques, and what support is available, will all help. Creating a farmer's group will help to make the best use of any information and support available.

- ⊙ Other sources of information include NGOs (Non-Governmental Organizations), schools, radios, books, universities and the internet.

- ⊙ Create a farmer's group, community group or seed saving group as a resource base. Other support groups, such as women's groups, are also very important. Representatives from each group can work together with representatives from larger groups, this will maximize the benefits of any information or support. In this way, all community members will receive benefits in the most sustainable way.

2.14 INTUITION

- ⊙ Using your intuition / instincts should be a part of your everyday decision making process.

- ⊙ Intuition is about sensing or instinctively knowing what to do and when to do it. These feelings come from trusting in yourself and from past experiences and knowledge from you, your family or your community. They also come from your spirit.

⦿ It is very important to look at all the facts and details, especially with technical work, but it is also important to follow your intuition. Intuition allows for more ideas, more creativity and more flexibility, and it will make each project more beautiful.

- *All plans and designs should be designed by the people who do the work. Any planning and designing should be done together and should include all the people who will be involved in the project.*

- *This means that women and children must be part of the planning process, especially for zone 1 and 2, where women do much of the work and children help a lot too. This will reduce the chances of mistakes being made and avoid wasting time, labour and expenses.*

- *Mistakes and changes will happen as you learn more and begin to use better techniques. Everyone makes mistakes, and by learning from these mistakes can make better plans in the future.*

- *Planning ahead will help us look to the future for our children and for the wellbeing of our nation.*

Permaculture Design Principles

SYMBOL	PRINCIPLE	DESCRIPTION	APPLICATIONS
	Observe and Interact Practice continuous and reciprocal interaction	Designs are created based on observation of and interaction with a given space's inputs and outputs. Careful observation and thoughtful interaction provide initial insight into existing working systems and ideas for potential future adjustments or redesigns.	Edible garden and food forest maintenance, redesign, evolution and transformation
	Catch & Store Energy Use existing natural capital	Designs include the capture and storage of natural resources (e.g. wind, sun, moisture, precipitation, geothermal heat). Designed systems mimic the self-sufficient systems found in plants and natural environments.	Wind towers, photovoltaic panels, heat pumps, solar ovens, grey water systems, passive solar building design, engineered wetlands
	Obtain a Yield Produce abundant natural and social capital and ensure regeneration	By creating regenerative systems and producing abundant yields permaculture creates long-term solutions, resource security and systems stability.	Edible gardens, food forests, keyhole gardens, bio-intensive gardening, agroforestry

	Apply Self-Regulation & Accept Feedback Understand positive and negative system feedbacks in order to reduce future management issues	While permaculture systems seek to mimic natural processes, they are still human designed systems that require maintenance for continued success.	Edible garden and food forest maintenance and management
	Use & Value Renewable Resources & Services Make use of and value existing, natural, renewable resources and services	Tools and processes exist throughout the natural environment that aid in the creation of resources and the minimization of waste – permaculture strives to better utilize and value these natural tools and processes.	Natural energy sources (e.g. sun, wind, water, biomass), human and animal labor, natural building materials (wood, adobe, cob, hay, rammed earth), biological services (cooperative microbial interactions, symbiosis)
	Produce No Waste Value frugality, and reuse "waste"	Permaculture systems are designed to make use of all respective inputs and outputs in order to minimize waste and pollution. "Waste" is viewed as a resource opportunity.	Recycling, material reuse and salvage, composting, sheet mulching
	Design from Patterns to Details Recognize natural patterns and design systems based on them	The principle of patterns seeks to mimic biological patterns for the purpose of system efficiency and self--sufficiency.	Earthworks (berms, swales, terraces), garden design and layout, biomimicry
	Integrate Rather than Segregate Connections between elements are as important as individual one	Inclusion and consideration of every element in a given design allows for the creation of symbiotic relationships that further promote regeneration, system stability, resource efficiency, and production of abundant yields.	Grey water systems, engineered wetlands, rooftop gardens, integrated production systems (e.g. aquaponics)

	Use Small & Slow Solutions Design systems to perform functions at the smallest scale possible; focus on self-- reliance, patience and reflection	Systems are designed first and foremost to function properly at the smallest scale possible in order to aid system efficiency and stability and future growth or replication.	Small scale permaculture systems and their respective growth and transformation towards larger scale systems

Source: Adapted from - Holmgren, David. Permaculture: Principles & Pathways Beyond Sustainability. Holmgren Design Services, 2002.

3

HOUSES, WATER & WASTE MANAGEMENT

3.1 INTRODUCTION: AN OVERVIEW

Everything is connected to everything else. This principle is very important to remember when creating sustainable agriculture systems. You can work with this principle for future benefits, or ignore it for future detriment. This principle is also applicable for the home and living area, including the kitchen, washing area and toilet

Every house is affected by the land and surrounding environment, for example by:

- Rainfall, erosion and flooding
- Wind
- Temperature
- The type of soil, rocks and trees
- Water availability
- Diseases (such as from mosquitoes)

To help reduce or prevent future problems, all these factors should be taken into consideration when building or renovating the house and living area.

Every house also affects its surrounding environment, for example by:

- Using food, firewood, cooking fuel, electricity, cleaning materials and other household needs
- Producing waste in the form of smoke, rubbish, waste water and human waste
- Farming practices

It is essential for the future to reduce our negative impact on the environment as much as possible. Here are some ways to achieve this:

- Reduce amount of polluting materials used, e.g. plastic bags
- Reduce pollution, e.g. affects of burning
- Reuse wastes, e.g. animal and human wastes
- Filter polluted materials from wastewater before the water returns to the river

Community ideas

There are many ideas for improving the quality of the home and living area which can be organized, applied and managed on a community level. Appropriate improvements will be better and less expensive if the community works together.

This can be achieved through community meetings and group agreement. It is important that everyone understands, gives inputs and has a sense of ownership of these community improvement projects. It is also important to work with the government, on a district and national level.

3.2 HOW TO CREATE A HEALTHY HOUSE?

- A healthy house is practical, long lasting, and makes life easier and better, while reducing costs. It is important to have a house you can be proud of, that looks beautiful on the inside as well as the outside. These considerations can all be combined.
- Making your house better does not necessarily have to cost more money. In fact, there are many ways to improve a house and living quality which will also provide extra resources for your garden and livestock, such as fertilizer for fruit trees, water for vegetable plots, and animal fodder.
- When you build a house or improve an existing house, it is important to take into consideration the following factors:

3.2.1 House Location

Build the house in a good location, take into consideration:

- The possibility of landslides
- The possibility of flooding
- From what direction strong wind comes
- The location of the closest water source
- How much sunlight there is
- Are there any trees to provide shade?

Sometimes there is not a lot of choice as to where you will build a house, but there are always many ways of reducing potential problems, which can help to create a better and more comfortable living area.

3.2.2 Ways to Reduce Risks

There are some things that can be done to reduce risks, such as:

- Stop erosion and reduce the risk of landslides. Above the house, swales can be built to catch water and soil. For this situation, direct the water slowly to one side, away from the house. This water can then be stored and reused for ponds below the house, compost pits or vegetable gardens. It is also important to plant strong trees right away to protect the soil and water

- Reduce the risk of flooding. Reforesting mountains and river banks is the best long term solution to reduce the risk of flooding. But sometimes flooding will still naturally occur

- Reduce the risk of fire. Fire will travel quickly uphill with the help of wind. If there is a high risk of fires, use firebreaks and other ways to stop fire on the path it could take

- Reduce the risk of house damage from strong winds. If possible, don't build on top of hills, and plant many trees to create windbreaks

3.3 COMMUNITY IDEA: PREVENTING DISASTERS

Preventing disasters is an issue for every family, community and nation. Rivers and river banks must be protected to reduce the risk of flooding. Planting trees, bamboo and grasses along the river bank will help reduce the risk of flooding and erosion. Reforesting community land will help reduce the risk of landslides.

Compatibility with Climate

There are many different kinds of climate conditions. A house should be designed to suit the climate of the area in which it is built.

Mountain areas can be very cold at night, so materials such as brick, rock or mud brick are the best to use as they will help keep the house warm at night.

Coastal areas are hot, so materials such as bamboo, wood paneling and grass thatch will help to keep the house much cooler than would cement and brick. An open house with an outdoor living area and good air flow will also help to keep the house more comfortable. Opening windows is also important. However, security is also an important consideration, so rooms that can be locked should also be built.

Mid-land areas (between the mountain and ocean) are best suited for combination houses, with rooms that will stay warm and other areas that are open. All tropical areas get hot, so outdoor shade structures can help to make the living area much more comfortable. Trees surrounding the outside of a house can also help to improve climate conditions by providing shade, reducing winds and cooling the air

3.4 GOOD HEALTH AND DISEASE PREVENTION

Much disease and illness can be prevented or the affects can be reduced by well designed and well built houses. This is especially true for the kitchen area.

PROBLEMS	SOLUTIONS
Smoke causes chest and breathing problems, which can cause TB (Tuberculosis)	Well ventilated kitchens
	Use smoke chimneys (pipes)
	Minimizing use of smoke producing stoves / ovens
	Don't use firewood
Mosquitoes cause malaria, dengue fever and many other diseases. They breed in still water	Don't let water lay in puddles / pools
	Cover all tanks and water containers
	Place mosquito netting on house windows
	Use mosquito netting when sleeping

Disease can spread because of an uncleanly / unsanitary washing area	A well built washing area is one that can easily be kept clean
	Use drainage systems that quickly drain wash water
	A simple filter system to clean washing water
Disease can spread because of dirty, open toilets	Use compost toilets and build toilets which prevent animals or insects from touching or eating human waste
	Use toilets instead of rivers
	Applying good toilet hygiene
Mice, dogs, cats, cockroaches, flies, etc can spread disease, especially if they eat food or manure	Protect all food in containers to prevent disease contamination
	Prevent animals from entering the kitchen
	Build a house that is easy to clean
	Wash hands before eating
Mould and damp walls can cause chest infection and breathing problems	Dry floors and living area
	A roof that is not damaged or leaky
	Keep rain away from the walls
	A well ventilated house
Burning rubbish produces smoke, which can cause health problems	Recycle rubbish, as much as possible
	Only burn rubbish in a specific area far from the house and children

3.5. A HOUSE THAT IS EASY TO CLEAN

A house that is easy to clean will reduce problems and improve health.

A cement or stone floor will be easier to clean. Walkways made from small or large stones can be built between the kitchen, house, washing area and toilet to prevent spreading mud and dirt, which will also reduce the spread of disease.

3.6. WASTE MANAGEMENT

Waste includes food scraps, used water, human manure and urine, plastic, paper, tin cans, bottles, smoke, ash, leaves, batteries, old car and bike parts, used oil, metals and many other types of waste. Waste also includes rubbish and pollution that is created when products are made and distributed, like foods, plastic toys, etc. Other types of wastes are created when we use energy, for example smoke from diesel generators. We contribute to the production of waste when we buy products and use energy. A well designed house reduces the amount of waste it produces. Being responsible for what you buy and use will benefit the future and help to protect the environment.

Following are some important guidelines that can be applied:

- Reduce waste that is produced
- Reuse or recycle as much as possible
- Be responsible about disposa

Good waste management means separating wastes and turning most of it into beneficial products, for example:

- Leaves are a valuable mulch material, which can be used to fertilize gardens
- Food scraps can be used as animal fodder
- Used water can be run through water trenches for use in the garden
- Compost waste water from the washing area by flowing it to banana trees
- Compost toilets turn human waste into fertilizer
- Wood ash can be used in compost and liquid compost
- Use plastic containers for storing seeds or seedlings
- Aqua bottles have many uses
- Tin cans can be used as seedling containers and watering cans
- Paper can be added to compost pits
- Used glass bottles can be cleaned and reused to store honey, oil, coconut oil, etc
- Old tires, cans, broken buckets, etc can be reused as seedling containers and flower pots
- And there are still many more examples

Bad waste management means burning everything, letting animals eat human waste, and leaving used water laying in puddles on the ground. And even worse waste management is dumping rubbish in the rivers. This causes pollution in rivers and oceans, which can create even greater problems in the future. It also looks ugly and spoils the beauty of our environment.

Burning Wastes

Some waste may still be burned. If rubbish, especially plastic, is burned at a very high heat it creates much less smoke, which is better for people and the environment. A circle of rocks around an area can create a place for rubbish to be burned. Put the rubbish into plastic bags and place in the burning area until there is enough rubbish to be burned all at once. This will help to create heat and reduce smoke pollution

Suggestions for the waste burning area:

- Keep the top covered and make sure everything is burned. This will also prevent dogs from entering
- Leave holes in the rocks to let air enter which will help speed up the burning process and keep the temperature high
- Position the burning area far from the house and not in the way of wind blowing towards houses

Don't let children stand near the smoke and breath it in. It is poisonous!

Community and Government Ideas

As soon as possible, build waste dumps in every village or district. This will greatly improve waste management, especially city waste. However, waste should always be separated and recycled first, then only the remaining rubbish be disposed of through waste dumps.

Examples of waste recycling:

- ⊙ Using old tires to make terraces. They are used in the same way as rocks are used to make swales / terraces. Trees can be planted below or even inside the used tires

- ⊙ Making compost from leaves, manure, etc

In small villages and communities where rubbish is burned, a community waste burning area can be made to reduce smoke and environmental problems. This area can be made of large rocks or coral, in the same way that it would be made for family use, only larger. Make a circle about 2 m wide and 1 m high.

Give extra attention and focus to types of waste, such as:

- ⊙ Toxic wastes, including used car oil, batteries, tires and medical waste

- ⊙ Glass bottles, tins

- ⊙ Scrap metal

- ⊙ Plastic

Recycling materials will slowly replace burning them. But remember, the best way to deal with waste is by not creating it in the first place! As a substitute, use natural, locally available materials whenever possible.

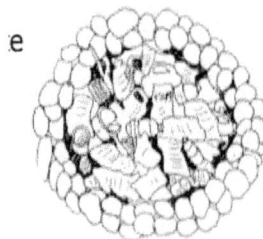

3.7 REDUCING WATER AND ENERGY USE

Water is a precious resource that often takes hard work to collect. This module offers many ideas for collecting and storing water, but first it is most important to reduce water usage. Even in cities and villages that have piped water, it is very important to only use what is needed, to ensure there will be enough water for use in the future.

Water saving ideas for in the home:

- ⊙ Always turn off taps after use

- ⊙ Make a compost toilet, it only uses small amounts of water

- ⊙ Reuse all kitchen and washing water for watering gardens

- ⊙ Use buckets and sinks for washing, and don't leave the water continuously flowing while washing

Energy is the fuel needed for a house. Wood, kerosene, electricity, generators, gas, petrol, oil and even candles all provide energy.

The price of oil and petrol is expensive and will only become more expensive, as supplies won't last forever. It is very important that people all over the world reduce energy use and change to using renewable energy. Some types of renewable energy are solar panels, micro-hydro generators, biogas generators and wind generators. (For more information about energy, see Module 12 - Appropriate Technology).

It is also important to use stoves and ovens that use only a small amount of firewood or none, or that use gas. Trees are being used very quickly, and are being cut down much faster than they are growing back. This is already a big problem for the environment, and it will only get worse unless changes are made. There are places in the world now where people have to walk all day just to collect firewood. Is this the future we hope for?

3.8 A LONG-LASTING HOUSE

Wood, bamboo, plywood and grass thatch are all natural materials that are comfortable and fairly inexpensive. However, they often only last four years, five years, or sometimes less, after which they need to be replaced. By choosing and treating the right material, especially bamboo, it will last a few years longer. Reducing rotting by keeping the ground dry will also help to increase the amount of time materials last. Stone or clay will last much longer than cement blocks. By covering the surface of stone, cement blocks, clay blocks and other building material with a render they will last much longer. Clay, sand, cement, cow manure, lime and water are examples of materials used in other countries. It is also common to cement block walls. Read further to learn more about these techniques.

3.8.1 Building a House

Include the whole family in the designing process so that all of their needs can be addressed. Women often spend a lot of time around the house and therefore will have many ideas about how needs can be met and how to deal with issues of health and cleanliness. This could include materials to use and ideas for making the home more beautiful. If these needs are met it will save time and labour, while improving the lifestyle of the whole family. For example, planting grape vines and passion fruit will provide shade and a cool area outside the house, as well as produce fruit for the family.

3.8.2 House Designs

Every region has its own designs for local traditional houses. These houses reflect the climate and available materials, as well as the taste of the people who live there. Brick houses generally are based on Portuguese designs, and more recently Indonesian designs. Sometimes the materials used are not suitable for climate conditions, especially for houses built in coastal areas. It is very important

- *A curved wall is stronger than a straight one, and more beautiful too!*
- *A house can be built with indoor and outdoor living areas*
- *Combine traditional house designs with modern house designs*
- *The position of the rooms is very important. A room that faces the afternoon sun will be the hottest room during the night. For example, a room that faces the afternoon sun is very appropriate for colder areas, but not appropriate for coastal areas*
- *A veranda or shade structure can be built at the west side of the house (where the sun sets) to help keep the house cooler during the night*
- *Building a house on stilts / poles will improve ventilation and reduce the risk of flooding*
- *A wide roof will reduce the amount of direct sunlight reaching the walls. This will help to cool down a house built in a hot area*
- *In areas with strong winds, a four sided roof is better than a two sided one for preventing wind damage. This is because the wind blows over the top instead of underneath, which can sometimes even blow off the roof!*

to choose building materials that suit the local climate. A houses shape and size will have impacts on its temperature, comfort, strength, durability and its resistance to disasters.

Future development and changes can also be planned, for example: If now you only build a small house because of limited materials or money, plan which rooms you would like to add on in the future.

3.8.3 Building Materials

The most common building materials used are wood, plywood, bamboo, grass thatch, brick, cement, tin and other metals. These are widely known and need no explanation of how to use them.

Bamboo, clay, stone and other traditional building materials have been used for a very long time, and are good building materials if used properly. Some ideas and techniques on how to use them will be explained only simply in this book.

3.8.4 Clay

If built properly a clay, mud brick or clay brick walled house can last for a very long time. In some countries there are clay or mud brick houses that are 100-200 years old, or more!

There are many areas that have good clay for making mud bricks or building clay walls. Clay houses stay much cooler in hot climates than do cement block houses, but good ventilation is still very important. Clay walls store heat throughout the day. At night, clay walls will slowly release this stored heat and help to keep the house warm. Clay houses are very suitable for areas that get colder at night.

Clay walls are made using clay, water and some dry grass. Mix all the materials together, then start building at the base and slowly build up.

Clay is more commonly used to make bricks. Clay bricks are made by combining clay and water (some dry grass may also be added). The material is placed into a mold and pressed, then the bricks are removed and dried. This process is similar to making cement bricks.

Two other type of blocks are stabilized earth blocks and adobe blocks. Stabilized earth blocks are made from clay, earth, and 10% cement. Adobe blocks are made from garden soil and grass cuttings.

SMART IDEAS!

- ◉ A render (sand and cement mixture) is essential for clay houses to last a long time. Using a small amount of cow manure in the render will help to prevent damage from insects and weather

- ◉ The roof should hang past the walls to protect them from damage due to heavy rains

3.8.5 Stone

Stone houses take a long time to build, but once they are built properly, they will last a very long time. Stone walls take a long time to heat up, so they stay cool throughout the day. Stone walls also store heat very well, so the house will stay warmer at night. Another benefit of using stone, is that it can easily be combined with other types of materials. There are many good stones / rocks to use, the main problem is transporting them.

3.8.6 Bamboo

Bamboo is a very well known material. It can be used for anything, including roofs, walls, decoration, furniture and much more.

Bamboo is very suitable for coastal areas because it has good ventilation. It is light, but strong and easy to work with. Selecting the right type of bamboo, cutting it in the right way and curing it will make the bamboo last much longer. (For more information on bamboo, see Module 8 – Forests, Tree Crops & Bamboo).

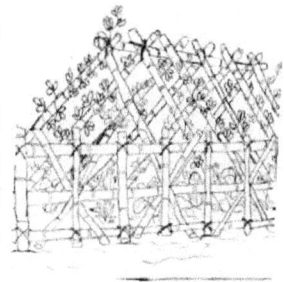

3.8.7 Combining Materials

Combining materials means building a house using different types of building materials. For example, using stone, wood and bamboo. This can be any combination of materials you like. A house built with different materials can maximize the benefits from each material used, for example:

- ◉ Clay and stone are the best materials for keeping a house cool during the day and warm at night. Cement blocks don't work very well, but are adequate if they are layered with a render

- ◉ Bamboo and wood provide good ventilation, and are fairly inexpensive

- ◉ Long lasting woods, like camphor and teak, make a good house frame / structure

- ◉ A tin roof lasts a long time and can be used to collect water. Grass thatching is inexpensive and gives good insulation. A bamboo roof can also be used to collect water

3.8.8 Ideas for House Improvement

These ideas can be used for building a new house, or renovating an already existing house. Simple, inexpensive improvements can make a big difference to the living area.

3.8.1.1 Ventilation

Good ventilation will keep a house cool in temperature. Hot air naturally rises. Air vents can be used to help the hot air rise and go out. An air vent is a small hole, approximately 30 cm x 15 cm in size, covered with wire screen to stop mosquitoes and mice from entering. If there are air vents near the roof, hot air can go out. If there are air vents near the ground, cool air can enter. When hot air leaves through the top air vents, cool air will enter through the lower vents. This is called convection. You need low and high air vents for convection to work. Opening windows will also help to cool down the house. Growing trees and plants around the outside of the house will make the air that enters even cooler.

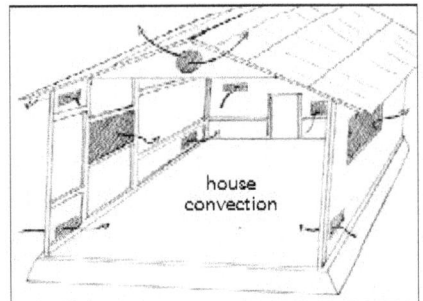

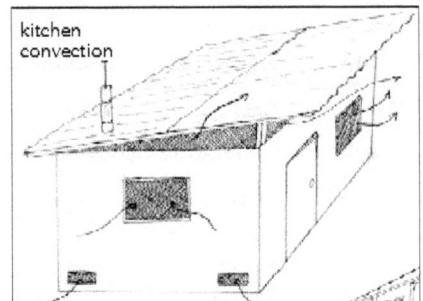

house convection

kitchen convection

3.8.1.2 Insulation

Render / Plaster

Render is a type of insulation that will keep the house cool during the day and warm at night. A render will cover and protect the walls, it can be used on cement blocks, clay, stone or even bamboo. For cement, clay and rock, the thicker the render the better the insulation, this will also help to ensure

sand

cement

dry cow manure

A cheap render mixture:

- ⦿ 8 parts sand
- ⦿ 1 part cement
- ⦿ 3 parts fermented cow manure (to make fermented cow manure, place fresh cow manure in a bucket with water, then leave for 5-7 days)

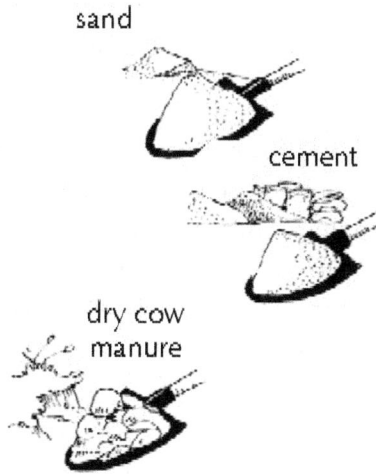

Another render mixture:

- ⦿ 1 part clay
- ⦿ 4 parts sand
- ⦿ 5 parts fresh cow manure
- ⦿ Hydrated lime (add water as needed)

Cow manure helps to seal the render and protect it from insects. Cow manure doesn't smell once it is dry! The lime helps to protect the walls from rain damage and acts as an insect repellent. wall lasts longer.

Render for bamboo and gedeg (woven bamboo walls)

This method works very well in colder areas, the process:

- ⦿ Cover the split bamboo / gedeg with chicken wire on the outside and inside, this will hold the render in place. Rendering both sides of the bamboo / gedeg will help protect it from insects and make the material last longer, while providing insulation
- ⦿ Apply the render until the wall is covered, till you cannot see the chicken wire or bamboo / gedeg. The thicker, the better

Curtains

In cold areas, curtains or cloth that covers the windows from the inside during the night will stop hot air from leaving through the windows and cold air from entering.

Roof Insulation

Traditional roof materials give very good insulation. Roof insulation will help to keep the whole house or building insulated. It is made to be placed below the roof and help keep the whole room cooler during the day and warmer at night. Insulating a roof can be expensive, but it gives comfort and saves money, because it will keep the room cooler you will spend less electricity on air conditioners or fans.

Natural Lighting

It is important to provide natural lighting in the house. If a room is too dark, it is harmful to the eyes, and you will need to use more lighting, such as candles and lamps. This is a problem for some types of traditional houses. Large windows in the house can provide natural lighting. If glass is too expensive or not available, close the window with wire screen to stop mosquitoes and animals from entering

Sky Lights

Sky lights can be used to increase the amount of light in a house. This can be a piece of clear plastic or clear sheeting that is placed on a part of a room's roofing.

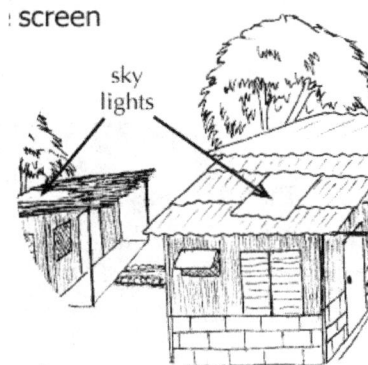

3.8.9 Making a House Last Longer

If you protect wood and bamboo from borers, termites and other insects, it will last a few years longer.

Bamboo (For more information on how to choose, cure and store bamboo, see the section 'Bamboo' in Module 8 - Forests, Tree Crops & Bamboo).

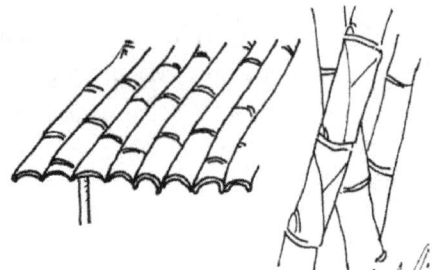

Wood

Wood can be cured by using:

- Paint or sump oil (old car or truck oil), especially on the base of a pole. This will help to protect the wood from termites and borers for much longer. Repaint the wood every 2 years to continue protection. Caution, this does increase the risk of the house catching fire

- Use a very hard wood, like eucalypt, to make the poles. White ants and borer insects will take a long time to eat the wood, or will not be able to at all

- Traditional ceremonial houses use wood that naturally lasts for many years. This knowledge can be acquired through the community, and this type of long lasting wood can be planted for future use

Cement post holders lift the bamboo pole off the ground, which prevents termites and white ants from eating the base of bamboo poles.

A simple method to make one:

- Dig a hole in the ground, with the same depth as would normally be used for a house pole

- Prepare an old bucket or used oil can to be used as the cement mold

- Fill the hole 50% with cement, place the pole inside the hole and pour in more cement until 10 cm below the hole's rim

- Place the bucket or can around the pole, fill with cement

- Remove the mold when the cement is dry

Keep the area and earth around the house as dry as possible. Damp earth around the house will bring white ants. They cannot live in dry earth. Damp earth will also encourage mold and fungus growth on walls, which can damage the walls and people's health.

Rendered walls will last much longer. Smoke from a wood fire will deter insects and dry out the walls, hence helping them last longer. However, too much smoke inside a house with a dry roof can be dangerous and harmful to people's health. Because of this, use a chimney or another way for smoke to go out.

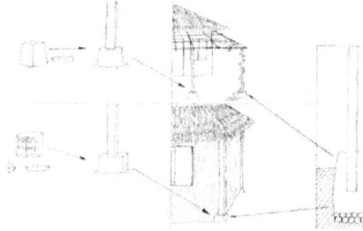

Roofing nails are much better to use on a roof than normal nails. They help to hold the roof on during strong winds and make it last longer.

Insect Screens

Mosquitoes cause many diseases. The risk of disease can be reduced by screening all windows and openings of a house. Use mosquito netting over beds for protection during the night.

3.8.10 Outside Improvements

3.8.10.1 Pergola / Shade Structure

A pergola or shade structure can be made large or small. This structure is simple to make and it provides shade for the outside living area, while keeping the house cooler inside. Different types of vines can be grown on the pergola, including passion fruit and grapes. Palm leaf can be used to cover the structure and provide shade until the vines grow over it. Pergolas can also be used to shade plant nurseries.

3.8.10.2 Trees and Windbreaks

Trees provide shade, protection from strong winds, and keep the house cooler because of moisture in their leaves.

Some trees grow too large to be planted near the house. Too much shade can cause moisture and ventilation problems inside the house. Also, the roots can damage the walls as the tree grows older. Be careful not to make too much shade over vegetable gardens.

In hot areas, a pergola / shade structure can be built, or trees can be planted, at the west side of the house (where the sun sets). This will help to keep the house much cooler during the night.

3.8.10.3 Gardens

Gardens around the house area add much beauty. Flowers, vegetables and herbs can be grown together. Trees and plants will help to keep the air much cooler because of the moisture in their leaves. Be careful not to

make gardens directly against wooden posts or walls, as this can cause rotting and insect problems.

3.8.10.4 Ponds

- ◉ Besides producing fish and vegetables, a pond adds beauty to the house area.

- ◉ A pond helps to keep the temperature cooler during the day and warmer at night. This is because water is slower than earth at increasing and decreasing temperature.

- ◉ Add neem leaves regularly to avoid mosquito breeding. Tilapia, gurami and mujair fish will eat mosquito larvae in the pond.

3.8.11 Kitchens

People spend a lot of time in the kitchen, because food is prepared and kept in the kitchen, so it is important to provide a healthy, clean and comfortable environment. The kitchen also includes a washing area.

A well designed and built kitchen should have:

- ◉ Enough ventilation – This is very important because kitchens often produce smoke

- ◉ Enough light – This is needed when preparing food

- ◉ Good sanitation, and should be easy to clean

- ◉ Good food preparation and storage facilities

- ◉ Clean water run-off trenches, so that water can be reused

- ◉ A stove and / or oven that reduces the amount of wood used and smoke produced

- ◉ A place to store and dry firewood

- ◉ No animals going in and out

An unhealthy kitchen is dark, smoky, difficult to clean, with water lying in puddles outside and animals going in and out. This will cause serious heath problems for a family and can spread disease. Women should be included in the process of designing the kitchen, because they understand and use the kitchen more than men.

3.8.11.1 Stoves and Ovens

A good stove and oven doesn't produce a lot of smoke. There are types of stoves and ovens that only use little or no firewood. (For more information about how to make and use good stoves and ovens, see Module 12 – Appropriate Technology).

3.8.11.2 Kitchen Ventilation

Ventilation is essential for reducing smoke in the kitchen. Smoke in the kitchen is one of the main causes of Tuberculosis (TB), and it can also cause many other health problems. Using plastic for starting fires is also dangerous because it is poisonous, especially in poorly ventilated kitchens. Even when kerosene or gas is used for cooking, good ventilation is still very important.

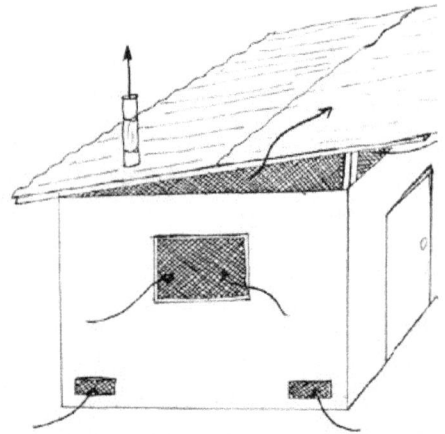

Types of ventilation could include:

- ⊙ Low and high air vents, which allow air to flow. Air vents are small holes (approximately 30 cm x 15 cm) covered with wire screen to prevent animals or insects from entering

- ⊙ Providing enough windows

- ⊙ A small gap between the walls and ceiling will allow smoke to flow out. Cover this gap with wire screen to prevent animals or insects from entering

- ⊙ A chimney can also be made to allow smoke to quickly flow out of the room

3.8.11.3 Enough Lighting

Dark kitchens are not good because they make food preparation difficult and can cause eye problems. Windows and skylights can be used to let more light in. Air vents will also help. Another solution could be preparing a separate outdoor food preparation area.

3.8.11.4 Sanitation

The kitchen and washing area is where many diseases are caught and spread. If the kitchen is healthy and easy to clean, than many diseases can be prevented. This does not mean that the kitchen should be cleaned more often, but that a well designed kitchen will make sanitation much easier to achieve and maintain. Vinegar and lemon are both good natural cleaning substances. Add a little to the water used for cleaning the food preparation area and floor, this will help to kill some bacteria which could cause disease.

3.8.11.5 The Food Preparation Area

It is best if the food preparation area is far off the ground, about waist height, and easy to clean.

This will:

- ⊙ Keep animals far away from the food

- ⊙ Make it easier to keep clean

- ⊙ Cause less back ache; staying in a squatting position for too long causes back pain and makes food preparation more difficult

3.8.11.6 Food Storage

Store food in bags or containers that animals and insects can't enter, like drums (for rice and corn), old biscuit tins, plastic boxes etc. Stop mice from getting into these containers.

3.8.11.7 The Floor

The kitchen floor is easier to clean if it is raised slightly higher than the ground outside. A floor made of rocks or cement are both good examples of materials which are good for health and can be easily cleaned.

3.8.11 8 Animals

- ⊙ No animals should be allowed to enter the kitchen. Chickens, dogs, cats and pigs all carry bacteria which can make people sick.

- ⊙ The risk of disease increases if animals are often around the kitchen area. Chickens, for example, often defecate in the kitchen, this is not just unhealthy, but it also brings a very unpleasant smell!

- ⊙ Save all food scraps in a bucket or container and feed them to the animals far away from the kitchen area.

⦿ Many kitchens have doors which prevent animals from entering. A door can be just a simple frame of wood with a wire screen. It costs a little money, but is worth it because people's health will improve.

⦿ Even just a low door to prevent dogs, pigs and chickens is good, but a complete door will also prevent mice and mosquitoes from entering.

3.8.11.9 Washing Area

It is healthier and easier to have the washing area higher above ground. Squatting for a long time is very hard on the body and bad for health. It is good to have a small table for kitchen dishes and tools to dry in the air, for example made of bamboo. Dish towels must be washed often because they easily become dirty and will spread bacteria from dish to dish. It is important to manage and reuse all water. Doing this will provide many benefits.

⦿ The following ideas may also be used for washing water and washroom water (if possible, combine them together to make less maintenance work):

  • Left over wash water can be run into a pond. Because this water still contains soaps and detergents, this pond should not be used for growing vegetables or fish. The water can be cleaned using water plants. These water plants will absorb the chemicals and nutrients from soaps and detergents and store them in their leaves. To clean household water properly for one family, the pond will need to be 3 m x 3 m or larger (about 1 m2 per person). First, fill the pond with sand and small rocks. Plant the water plants in the sand and trim regularly, this plant trimming may be used as mulch for fruit trees. A place for water to overflow must be made, especially for in the wet season. This overflow water can be run through trenches into a compost pit or through vegetable plots.

  • A trench can be made by digging about 40-50 cm deep, about 5-10 m long, and 50 cm wide. Fill the trench halfway with sand and small rocks and grow water plants to improve the trenches water cleaning ability. Any extra water that flows from the trench can be run into a compost pit or through swales. Banana and papaya trees can also be planted along the edge of the trench.

- On sloped land, water can be run through a pipe or small trench. For areas where water is in shortage, this method works very well and is practical and easy to maintain.

SMART IDEAS!

- ◉ Any ponds, trenches or swales which are used to clean water from the kitchen, washing area or bathroom should have a small fence built around them to prevent animals, like ducks, pigs, goats, cows and buffaloes, from drinking the water or eating the plants. This unclean water can make animals sick and animals drinking the water could damage the cleaning system

- ◉ Water hyacinth is a good plant to use for cleaning water. It lives on top of the water and multiplies very quickly. Lotus plants will also work well

Many people wash clothes in the river because there is no other water supply available. When water does become available in villages, it is better to stop washing clothes in the river to reduce pollution from washing powders and detergents. This will also save time and energy. Create a washing area that can be combined with dish washing and the washroom so that water can be easily collected, cleaned and reused for watering fruit trees and vegetable plots

3.8.12 Washrooms

There are many different ways to make a washroom, from simple compost showers to washrooms made of rock, clay or cement blocks. Choose building materials that are most easily accessible to you. The most important thing is to reuse the left over water!

3.8.12.1 Compost Showers

Compost showers are made by digging a hole about 2 m wide and 1 m deep. This is a inexpensive and simple way to directly reuse water

There are 2 methods of making a compost shower:

Method 1

To allow water to easily be filtered into the earth, fill the hole with:

- ◉ A 10 cm layer of palm fiber, this should also cover the sides of the hole

- ◉ Add rocks / coral / stone until almost full

- ◉ Add gravel to fill in the gaps, and a top layer of gravel about 5-10 cm thick

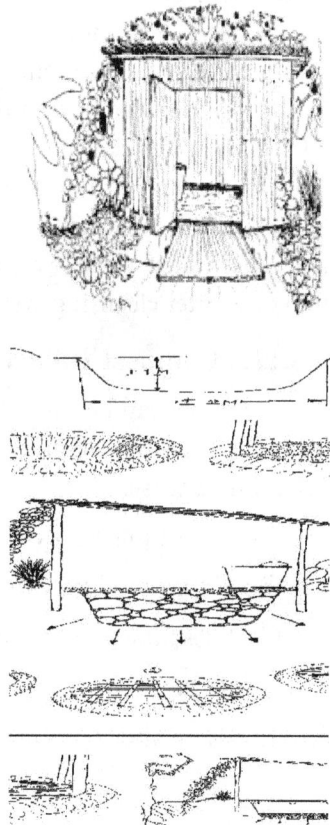

Method 2

Fill the hole half way with coffee or rice husk. This will soak up the water. Then, make a floor with eucalyptus poles and cover with woven bamboo panels. This method will rot within 1 or 2 years, so the floor will have to be made over again.

Build a simple structure surrounding the shower to give privacy and to provide a trellis for vines to grow. All the water will be stored in the ground to be reused by trees and plants. Plant banana trees, pumpkins, gourds, luffa, watermelon, papaya trees, pineapples, chillies, tomatoes, passion fruit and other plants around the edge of the shower, these plants will absorb and reuse the water coming from the shower.

3.8.12.2 Washroom Designs

A washroom can be made from any available materials. Clay, rock or cement will last a long time. Use a cement based render or tiles to protect the walls. A rock, cement or tile floor is easy to keep clean and hygienic. The washroom doesn't have to be square! Bamboo, wood or woven bamboo panels can also be used, but these materials will grow moss and become moldy because of moisture so they will need to be replaced more often. The mold is also a health risk and can spread disease. If you use a tin roof, water can easily be collected in a drum or tank for use in the washing area. Used water can be run out through pipes or trenches to be reused. If possible, use a pipe because it will be easier to maintain. Look in the kitchen section for ideas on how to clean and reuse water. You can also combine the used water from the washing area and kitchen into one water cleansing system.

3.8.12.3 Compost Toilets

Human waste can be turned into valuable fertilizer. But it must first be treated and composted properly to avoid spreading disease.

A compost toilet provides many benefits:

- ⊙ Makes good fertilizer
- ⊙ Uses little or no water
- ⊙ Reduces and prevents disease problems

This is turning a problem into a solution.

3.8.12.4 Compost Pit Toilet

A compost pit toilet is very simple to make and use.

Dig a large hole, about 1-1.5 m deep and 2 m in diameter. Use the dug out soil to make a mound around the edge of the hole.

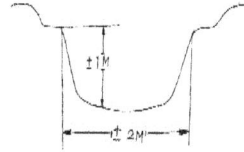

Make a strong floor using a board to cover the hole. Make a small hole in the center of the floor, this will be the toilet hole. Make a lid to cover the hole for when the toilet is not in use.

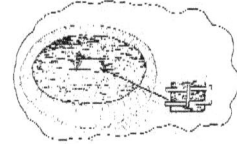

Build a simple toilet house around the toilet pit to provide privacy. This can be made of wood, bamboo, palm leaf or grass. Use whatever materials are most inexpensive and easily available. Build the door in the direction where the wind most often comes from. This will help to reduce bad smells.

A ventilation pipe can be added to the toilet hole to increase the speed of composting and to reduce any bad smells. This could be a bamboo pipe, with the insides cleaned out. Insert the bamboo into the hole in the toilet floor. To prevent flies or insects from entering, make sure any gaps between the pipe and floor are sealed.

The time it takes to become full depends on the size of the pit. Usually it will take around 1-2 years. When the toilet is full, dig another pit to be used. Add more leaves, rice husks and other materials to the first pit and leave it to compost for at least another 6 months. After this time, the manure can then be removed and used around fruit trees. The pit can then be used again. By this time the floors and walls will need to be rebuilt.

How to use a compost pit toilet:

- ⊙ Add one large handful of rice husks, coffee husks or sawdust every time the toilet is used. This will turn the waste into fertilizer and stop it from smelling. This is very important! Add one bottle of EM (Effective Microorganisms) every month to help the composting process

- ⊙ Add about 5 handfuls of wood ash or lime every week. This will help the manure decompose faster and make better quality fertilizer

- ⊙ Always keep a lid on the toilet hole in the floor when the toilet is not being used. This will prevent flies from entering the hole. Flies can spread diseases from the manure

- ⊙ No water is needed. The manure works better with little or no water added. It is better not to use the pit for urinating. Urine can instead be used to fertilize mature fruit trees

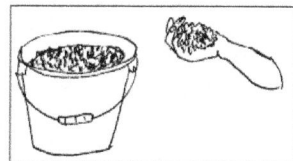

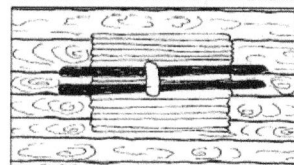

⊙ Dig the pit as far away from the river as possible. This is because in the wet season bacteria from the toilet might enter into the river through ground water. This can cause disease if people use the contaminated river water

3.8.12.5 Plants

Banana, pumpkin, luffa and passion fruit are the best plants to grow around the edge of the toilet. Citrus trees can also be planted nearby. The bad bacteria (and taste) doesn't transfer to the plants or fruits. Don't plant root vegetables that might directly touch the decomposing manure, this could spread disease.

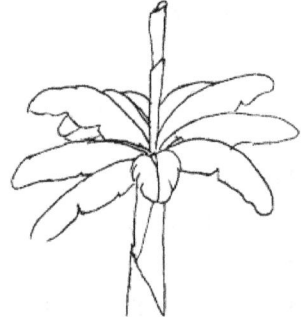

3.8.12.6 Compost Toilet Systems

There are many different types of compost toilets with many different designs. If you are interested, there are lots of internet sites and books that offer detailed information about these different types of toilets. In this book, only one type will be explained, it is a simple design that is easy and inexpensive to maintain, and is already being used in many countries.

3.8.12.7 Two Box Compost Toilet

This compost toilet is simply two cement boxes joined together.

The boxes are made of cement blocks. Each box is about 1 m3 (1 m x 1 m x 1 m) on the inside. The cement blocks must be rendered on the inside and outside to make them waterproof. On top of the boxes, make a cement slab about 10 cm thick.

Use steel reinforcing rods in the cement to strengthen it. This is important because the toilet must be strong enough to support the weight of the people on top.

On top of the boxes there is a toilet hole, about 20 cm wide, in the middle of each box. Each hole needs a lid that fits well for when the toilet is not in use. Each box has a small door on one side. This is to remove the compost when it is ready to be used. The door should be big enough for a shovel to fit through.

A simple wall should be built around the toilet to provide privacy. The easiest method is to add wooden or bamboo poles at each corner while the cement boxes are still being made (while the cement is still wet).

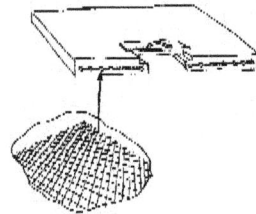

Remember to make the door to the toilet facing where the wind most often comes from, this will help to reduce any bad smells.

3.8.12.9 Ventilation pipe

A ventilation pipe improves composting and reduces bad smells. They are used for most types of compost toilets. With this type it is not essential, but recommended. Use a piece of bamboo or a pipe about 1.5 to 2 m long. This air ventilation pipe should be attached while still in the process of making the top boxes (while the cement is still wet) so that the bottom of the pipe is inside the box. Air is then drawn out from the container through the pipe.

3.8.12.10 Water conservation

This toilet does not need any water. In fact, it will not work if water is used. Instead, use toilet paper (tissue). Water will flood the system and stops the manure from decomposing. People using this toilet should urinate elsewhere, because too much urine can also cause problems. Urine can be mixed with water and used for other things, like fertilizing fruit trees.

Removal pipe

A removal pipe can be added at the bottom of each box. This will allow any excess liquids to flow out. Wire mesh must be added at the start of the pipe to prevent any solid items from also flowing out. When adding a removal pipe, there are some important factors to consider:

- The wire mesh may often become blocked and will need cleaning
- The liquid that comes out must flow through a water cleaning pond or trench, such as described in the kitchen section

The way to use this toilet is the same as using a compost pit toilet.

For one family it will take about six months to completely fill one box. At that time, the other box can be used. Leave the first box for six months so that the manure can decompose and become compost. There is no need to stir. After the second box is also full, the compost can be removed from the first box and it can then be used again. This compost is high quality and is good for use on fruit trees, but is not recommended for vegetables because it is too strong.

SMART IDEAS!

- If there are too many people using the toilets, it is better to build more boxes than to make the boxes bigger. It is better to have enough boxes so that the material in each box can be left for six months to decompose and become compost.

This system takes more time and money to construct, but it works very well if maintained properly. There are many different types of compost toilets that can have urine and small amounts of water go through them, however toilets like these need removal pipes which can remove liquid from the toilet box and flow it directly into a water cleansing system. These toilets are good for large houses, and especially for ecotourism, offices and towns.

More research must be done before any attempt to make a compost toilet. If it is not built correctly, it will require a lot of extra maintenance and will not produce good compost. Most houses in the city have septic tanks. Using these septic tanks will reduce bacteria problems, and therefore reduce disease caused from this bacteria.

SMART IDEAS!

- ⊙ In mountain areas where it is very cold at night, drops of dew can be collected using metal roofs. This dew water can be stored in drums or storage tanks, especially during the dry season. Even though this is just a small amount of water, it will still reduce the hard work of collecting water

- ⊙ Store water in the wet season. Rain water can be stored in tanks, but usually tanks will not hold all of it. The extra water can be stored in the ground, in ponds, and by trees which you can plant (trees store water in their roots, trunk, branches, and leaves). A shallow trench can be dug around the house in places where the rain falls, and then filled with gravel. Use the soil from digging the trench to make the ground higher on the side closer to the house. This will help to keep the house dry during the wet season. Design the trench so that water flows away from the house. Water can be run to vegetable gardens, compost pits, etc

Some important things to consider:

- ⊙ Build septic tanks as far away as possible from wells, water pumps and rivers. Overflow from septic tanks can pollute water supplies and can make people sick

⊙ Add a small amount of lime twice a year to help balance the pH levels. pH is the measure of acidity or alkalinity. (For more information about pH, see Module 4 - Healthy Soil)

⊙ Don't use bleaches for cleaning the toilet, because it kills good bacteria needed for decomposing manure

3.8.13 Water Supply and Storage

3.8.13.1 Collecting Water

Collecting water is hard work, which takes up many hours each day. Women and children most often do this task. If water can be collected closer to the house, much time and energy can be saved and used for other activities. This will improve the life of the whole family.

3.8.13.2 Family / Household Water Collection

The roof of the house, kitchen and washroom can be used to collect water. A tin metal roof can catch a lot of water when it rains. Bamboo cut in half can be used as a gutter to collect water and flow it into a tank or drum.

Water can be brought into the house through metal, plastic or simple bamboo piping.

Water for gardens and ponds can be collected through swales. Make the swale trenches so that the water flows slowly in one direction. At the end of where the water flows, redirect the overflow water using rocks. This water can be directed into a pond or water storage hole, to be used for animal, vegetable and fish production.

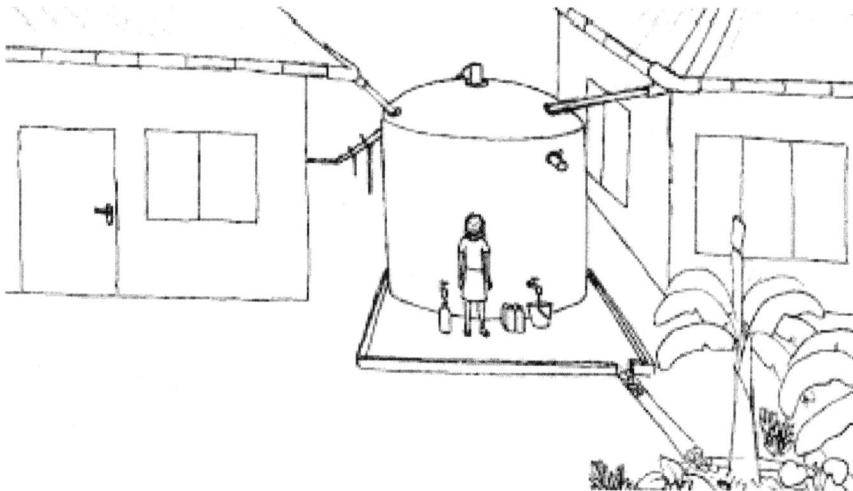

Community Water Collection

- Water springs are a traditional source of water. These springs need to be protected from animals and damage. An animal drinking hole should be separate. Bamboo or metal pipes can be used to run water to communities, which can then be stored in large permanent storage tanks. Overflow water from the storage tanks can be used to water fruit trees and vegetables, using pits or swales. Using the overflow water in this way will also reduce mosquito problems

- Water pumps and bores are other good ways to collect water from places close to the house. They can be made for each house, but will be much more inexpensive if they are used for a group of houses or communities

- Community wells. A community well must be kept clean to avoid the spread of disease through dirty water. The well should be built with a circular edge going up 1 meter high, this can be made from stone and cement mixture. This will prevent dirty water entering the well or animals making the well dirty. An animal drinking hole should be separate. Don't use dirty buckets or cans as water containers. Make a cover for the well to help reduce mosquito breeding

- Water can be pumped up hill to be stored in tanks, petrol pumps or ram pumps can be used. Ram pumps do not use petrol and require much less maintenance. Ram pumps can work very successfully. (For more information about ram pumps, see Module 12 – Appropriate Technology)

- Working with the government to provide water for cities or villages.

SMART IDEAS!

- In mountain areas where it is very cold at night, drops of dew can be collected using metal roofs. This dew water can be stored in drums or storage tanks, especially during the dry season. Even though this is just a small amount of water, it will still reduce the hard work of collecting water

- Store water in the wet season. Rain water can be stored in tanks, but usually tanks will not hold all of it. The extra water can be stored in the ground, in ponds, and by trees which you can plant (trees store water in their roots, trunk, branches, and leaves). A shallow trench can be dug around the house in places where the rain falls, and then filled with gravel. Use the soil from digging the trench to make the ground higher on the side closer to the house. This will help to keep the house dry during the wet season. Design the trench so that water flows away from the house. Water can be run to vegetable gardens, compost pits, etc

3.8.13.3 Water Storage

Storage tanks with taps can be used to store water. This makes water usage much easier. A storage tank can catch water from water pipes, pumped water, or even water collected through a roof. This tank can be made of cement, plastic, tin or aluminium. Old drums can also be used, especially for catching water from the roof. These drums must be cleaned well to remove all the left over petrol. How to do it:

- ◉ Wash well with detergent

- ◉ Rinse with water

- ◉ Dry out for one week in the sun before use

3.8.14 Reforestation Around the Village

Reforestation around villages and cities will help very much to store water, because:

- ◉ More rain water soaks into the ground, reducing erosion

- ◉ It will keep the store of ground water more constant. This is very important for future water supply

- ◉ It will provide leaves for mulch, which will also keep more water stored in the ground

- ◉ It reduces strong winds, which can dry out the soil

This is very important, especially for places where communities are collecting water directly from the ground water. Without trees, water quality will drop, and water levels will also drop, making it more difficult to reach this water. This has already happened in many countries.

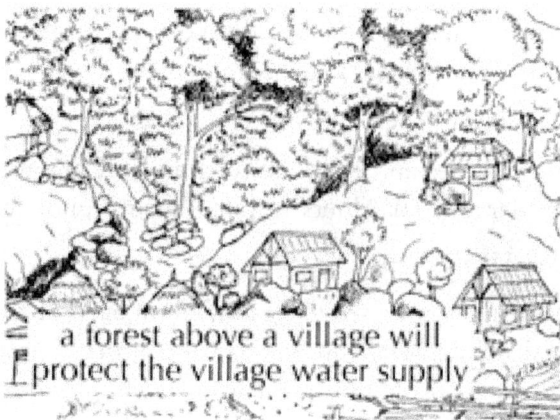

a forest above a village will protect the village water supply

3.8.15 Keeping Water Clean

House water that is stored must be kept clean. This will reduce the chances of disease.

This can be done by:

- ◉ Covering water storage tanks to prevent mosquito breeding

- ◉ Not using dirty buckets or cans to carry water, especially if you do not use taps

- ◉ Use moringa seeds, detailed instructions will follow

- ◉ Regularly clean water sources and piping

3.8.15.1 Cleaning Drinking Water

Moringa Seeds

The seeds of the moringa tree can be used to clean water of dirt and most bacteria. This is a simple and effective way to make water drinkable. It is used in Africa, India and other countries. This technique also saves a lot of fire wood, energy and time because water does not have to be boiled.

How to use the moringa seeds:

- ◉ Remove some seeds from the pod and peel the outer shell off the seeds
- ◉ Crush the seeds into a fine powder. Don't use discolored seeds (brown colored)
- ◉ Add 2 small spoons of this moringa seed powder into 1 clean water bottle (1500 ml aqua bottle)
- ◉ Shake for 5 minutes
- ◉ Filter this solution through a clean cloth into the bucket of water that is to be treated
- ◉ Stir quickly for 2 minutes, then slowly for about 10 minutes
- ◉ Leave still for 1 hour. The dirt and bacteria (usually between 90%- 99%) will stick to the moringa seed powder and sink to the bottom of the water
- ◉ Carefully pour the clean drinking water into clean bottles or containers, leaving the powder at the bottom of the bucket

Community ideas

You can use moringa seeds to clean large amounts of water. Use about 1 kg of the seed powder for 10,000 liters of water (about 1 gram for every 10 liters).

Clay Water Filters

Water filters are used to clean water of bacteria that can make people ill. This means that all drinking water can be cleaned and that the water does not need to be boiled before drinking. This filter is made using bowls of clay which have been fired in an oven. These bowls have a special base which is different than the rest of the bowl. Water can pass slowly through this base.

How do clay water filters work?

Clay is made up of millions of very small particles which are joined together. The clay particles when joined together form even smaller holes between them. Water will pass through these holes very slowly. The size of the holes depends on what type of clay is used, different types of clay have different sized particles and different sized holes between them. Unclean water carries a lot of bacteria, but the size of the bacteria is larger than these holes at the base of the clay bowls. As the water flows through the holes in the clay, the bacteria becomes separated from the water. This makes the water safe enough to drink.

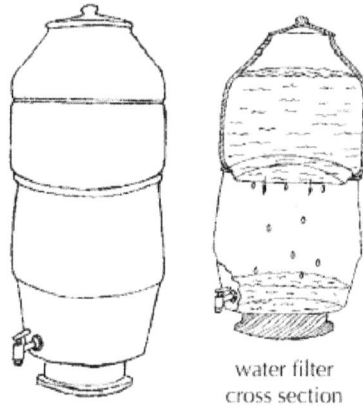

water filter
cross section

These clay water filters must be fired in a proper oven to work well. The temperature of the firing also affects the size of the holes, because clay shrinks as it is fired, so the holes become even smaller. The effect can be as follows:

⊙ If the holes in the clay are too small, the water will take a long time to flow through

⊙ If the holes in the clay are too large, bacteria will not be stopped and the water will not be clean

This is why meticulous testing and proper firing must be done so that clay filters work well.

How do you use a clay water filter?

This type of water filter is very easy to use. Simply pour water into the top bowl, the water will slowly filter through the base into the bowl below it. Bacteria and dirt will be caught and will stay in the bowl above. Drink the water from the bottom container. The base of the top bowl may become blocked from bacteria or other material that becomes caught, so it should be cleaned often to continue to work well. Use a brush, boiling water, lemon and vinegar to clean it, do not use washing powder. There are many other types of water filters. One other type is very similar, but instead it uses silver nitrate (a natural antibacterial material) in the top clay bowl to help clean the water.

3.8.16 Reducing Mosquito Problems

Extra water during the wet season can be stored in ponds or banana pits by using trenches and swales. This will prevent stagnant water forming on the ground, so that mosquitoes will only be able to lay their eggs inside the pond. Fish, frogs, lizards and insects that live in and around the pond will eat the mosquito eggs and larva in the water. Because of this mosquitoes will reduce in number, and hence the risk mosquito spread disease will also be reduced. These insects and small creatures will also eat some of the pests that feed on your vegetables.

Other ways of handling mosquito problems:

- ◉ Don't leave water stagnant in open places, cover all water tanks and containers
- ◉ Treat waste water properly
- ◉ Keep small fish (tilapia are best) in containers of water that will be used for washing, not in water that will be used for drinking, they will eat mosquito larva
- ◉ Keep fish in rice paddies to eat mosquito larva
- ◉ Prune off old banana leaves regularly, because mosquitoes like to stay there
- ◉ Add neem leaf regularly to every pond
- ◉ For compost pits, soak a handful of neem leaves in a bucket for 2 days, then pour some liquid with the neem leaves into every compost pit. Repeat this every 3 months

Mosquitoes represent not only a community issue, but also a national issue. Education about disease prevention and mosquito life-cycles is very important. Keeping the community water supply free of mosquitoes will help to reduce this problem.

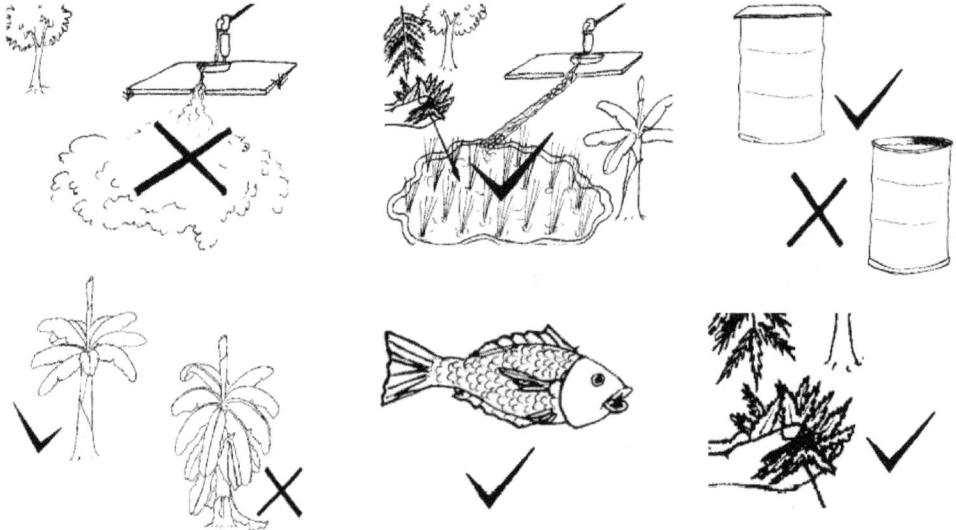

3.9 COMMUNITY BUILDINGS AND LAND

Community buildings and land are a great opportunity to give examples of how to improve the community. This could include:

- ◉ Examples of how to improve housing
- ◉ The community making compost toilets, on community land for community use

- Collecting and storing water using community buildings
- Examples of stoves, ovens, and other appropriate technology
- Combine these ideas with other ones, like gardens, nurseries, seed banks, etc
- Working with schools and community organizations as part of the community development process

3.9.1 National Plans

National plans for improving community housing is very important. This includes:

- Clean water supply
- Waste management
- Disaster prevention
- Education about health and disease prevention

Communities also need to work together with the government to develop plans, but first the government must hear from the communities what they most need and how they can work together in achieving this.

3.9.2 A Continual Process

Housing and water supply is a continual process. Improvements can always be made more beautiful while saving time and by using different technologies to achieve more benefits.

4
HEALTHY SOIL

4.1 INTRODUCTION: AN OVERVIEW

Healthy living soil is the foundation of any farming activity. Soil is the most important factor in producing healthy and productive vegetables, fruits and grains. Soil must contain all the nutrients that are important for plant growth. The soil must be protected from erosion to keep a good top soil, and protected from the sun and wind to conserve its moisture.

Creatures / biota in the soil must be protected because they are essential for creating healthy, living soil.

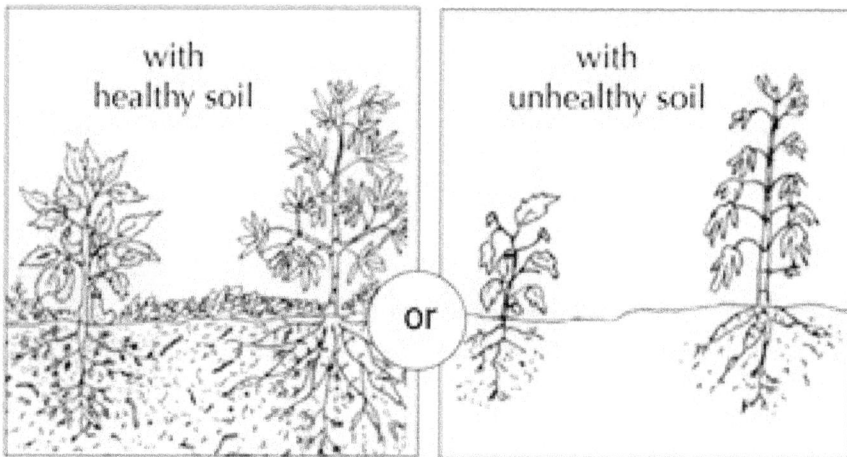

Good quality soil is very important in all gardens, small and large. The whole family, especially women who do most of the home gardening to supply nutrition for the family, should learn and understand about soil quality and techniques for improving soil quality. Most of these techniques are simple, do not require heavy work, and use local, inexpensive materials.

Better quality soil will give better quality produce, with better nutrient supply and better taste. This is a good way to directly improve family health. Better health reduces the chances of becoming sick, increases thought and concentration, gives strength, energy and a longer life. Good quality vegetables make people feel full when they eat them, and keep them full for longer.

In some places, the earth's top soil and nutrients are slowly disappearing because of regular burning and deforestation of farming land and forests. This causes erosion and landslides. These practices must be stopped! Farmers must protect their soil. The soil should be viewed as a very valuable asset.

4.2 HEALTHY LIVING SOIL

- ⊙ Healthy soil contains humus. Humus is partly broken down organic matter: compost, mulch, manure, plant roots and plant material. Humus provides food for soil biota, which then become food for plants. Humus also stores plant nutrients, helps to bind soil particles together, improves soil structure, and soaks and stores water in the soil

- ⊙ Healthy soil means that the soil is alive! It contains millions of soil biota which turn organic matter and nutrients into plant food. Soil biota includes bacteria, microorganisms, ants, worms, and many other very small organisms

- ⊙ It contains a balanced combination of clay and sand particles. The clay holds the minerals and the sand allows drainage / water channels

- ⊙ It is composed of 50% clay, sand, humus and organic materials and 50% air pockets. The texture should be loose when pressed, not crumbly like sand or sticky like clay

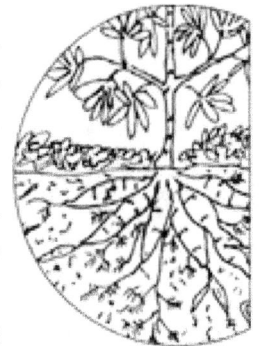

Air pockets are very important because:

- They provide space for the soil to hold a lot of water
- The air provides the oxygen that is needed by plant roots to process nutrients
- They allow easy, fast and deep root growth, so that the plants can soak up more water and nutrients, and the plants will become bigger and healthier
- The soil can function as a 'nutrient bank', it stores nutrients that are ready for plant use, and those nutrients are then not lost from the soil
- The soil will have a balanced pH level. This means that the soil is not too acidic and not too alkaline

4.2.1 The Importance of Worms in the Soil

Worms are your best friends in the soil!

These worms are earthworms. This is a different type of worm than the ones that make animals and people sick. Many worms in your soil show that the soil is healthy. Earthworms eat the humus in the soil, and then change that humus into nutrients, this is very good for the soil.

Earthworms will continuously:

- Change humus into nutrients that plants can use
- Dig the soil so that air can enter the soil
- Improve soil structure and water drainage
- Bring nutrients up from deep in the soil to supply food for plant roots

All the worms need is mulch and compost! However, be careful with chemical pesticides, herbicides and some fertilizers, because they will kill the worms in the soil.

4.2.2 Benefits of Healthy Living Soil

- Plants are more drought resistant because the soil can store much more water and plants can send their roots much deeper into the soil to receive water and nutrients
- Plants are more disease and pest resistant because they are healthier. An unhealthy person will become sick more often, the same is true for plants
- The plants produced will contain more vitamins and minerals, which if consumed will improve the health of the whole family, especially children

- Reduces evaporation from the soil, so that the soil will hold and store much more water. This will reduce the need to water plants

- You have millions of workers in the soil that manage nutrient availability, store those nutrients, and increase the amount of air in the soil. Worms are hard workers!

- The soil becomes easier to dig and work with because it has a loose texture. This is very important because it will save a lot of time and human energy

- It can save a lot of money if most of the land management is organic. Soil needs very little expense if good techniques are used. Remember to compile and reuse all plant and animal wastes

- Water will not be stagnant in the soil during the wet season. Even though healthy soil can store more water, the good soil structure will also allow for drainage if there is too much heavy rain. Too much water can slow down plant growth, and even kill plants if their roots become drowned in water. In areas where the soil contains too much clay, stagnant water can become a big problem. Making raised garden plots will also greatly reduce this problem

To improve soil, do:

- Use organic compost, mulch and EM (Effective Microorganisms) regularly. This will provide a lot of nutrients, increase the amount soil biota, improve soil structure and they are inexpensive to make

- Use mulch to protect the soil from direct sunlight, conserve water and increase the amount of humus in the soil

- Recycle organic materials, such as left over plant and animal material, to return nutrients into the soil

- Use legumes. There are many different types of legumes that can be planted, from seasonal to perennial. Legume plants provide nitrogen for the soil, can be used for mulch, animal feed, food for people, serve as windbreaks, help to prevent erosion, and more

- Rotate crop production. Different types of plants need different types of nutrients. Crop rotation is useful for balancing nutrients in the soil. Crop integration will also help

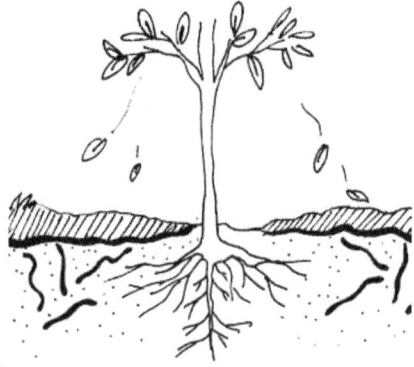

To protect soil quality, don't:

- Compact the soil. Soil compaction reduces root growth, water storage and water drainage, as well as damages soil structure. It also means that a lot of energy is needed to dig the hard soil

- Leave the soil open, exposed to the sun. This will make the soil dry and more difficult to dig

- Use anything that will kill soil biota. Soil biota are your friends and helpers for building healthy and balanced soil. Using pesticides and herbicides will kill them

- Waste water. Water is a precious resource and should be stored in the ground. Water that is continually flowing can create erosion. Good water usage will reduce the risk of drought. The amount soil biota will also reduce if the soil is very dry, these biota need water too

BEWARE!

Stop erosion...

The first soil that is eroded is the topsoil. This is the most valuable layer of soil! The topsoil contains a lot of nutrients that could take years to replace. The soil will not be able to hold water, and plant roots will become exposed to the soil surface, the plants will then grow very slowly or even die.

Stop burning...

- Burning destroys valuable materials, that can be made into compost, mulch, and nutrients for the soil

- Burning reduces the amount of soil biota

- Burning dries out the soil and reduces water volume

Burning creates erosion and pollution

4.2.3 Different Types of Soil

By doing a simple experiment, you can identify the types of soil that you have. This knowledge will help you in choosing the best method for improving your soil.

- First, take three or more soil samples and place them in clear jars or bottles
- Fill the container 2/3 with soil, then add water until full
- Close the containers and shake them evenly
- Then, let the soil settle and you can see what type of soil you have

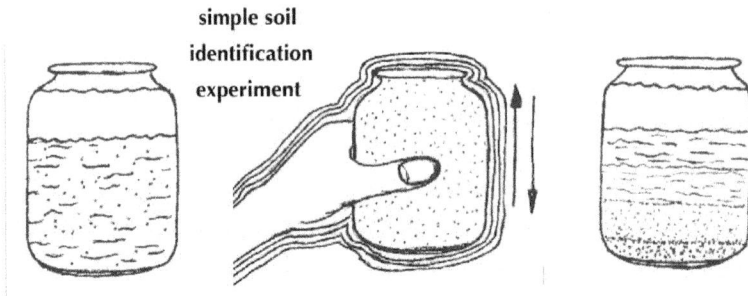

simple soil identification experiment

Clay will always be at the top, with sand underneath, and very course sand at the bottom. This is a very simple experiment, so even kids can do it.

Clay soil holds nutrients well, but does not contain much air, so when heavy rains come the water can become stuck in the soil.

While sandy soil will soak up water quickly and contains a lot of air, it easily releases nutrients and can quickly become dry.

4.2.4 Improving Soil Quality

For All Types of Soil

- For all types of soil the best solution is to regularly use mulch, dry compost and liquid compost. This will:
- Improve soil structure and the amount of air in the soil
- Increase the number of soil biota
- Increase the amount of available nutrients
- Increase water storage capacity

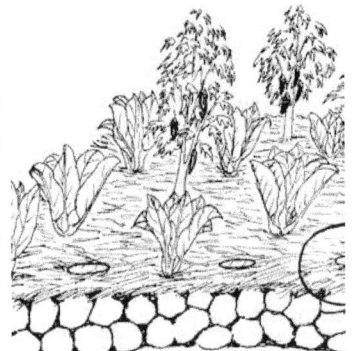

For Clay Soils

The following steps are useful for improving clay soils:

- Reduce compaction, because once the soil becomes compacted it sticks together. This makes root growth difficult, as well as making it difficult for people to dig

- Add sand to improve soil structure

- Use green manure crops and crop rotation to help improve soil structure over time. See the section on legumes in this module for more information on techniques

- Planting trees will also help to improve the structure of clay soils. Trees provide mulch material and their roots will help to break up the clay soil. Trees can also be combined with other types of plants

- Gypsum can help to improve the drainage and structure of soil. This technique will improve clay soil structure quickly, but is expensive. This technique will not work well if the soil's pH is too alkaline

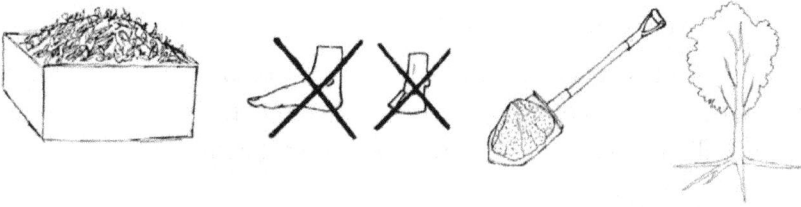

For Sandy Soils

The following steps are useful for improving sandy soils:

- Add 3 shovels of clay into liquid compost. The clay will bind nutrients, and when this mixture is used, the clay will stay in the sandy soil and hold nutrients within the soil

  adding clay to liquid compost

- Add 1/2 a shovel of clay to a large bucket of water, spray this mixture over the sandy soil. Using the liquid compost technique above is much better, but this method still adds valuable clay particles to the sandy soil

- Use green manure crops to add humus to the soil, this will improve the sandy soils structure

- Plant trees. In dry sandy areas, it is better to plant trees than to plant annual vegetable crops

Soil pH

The soils pH level is a measure of the acidity or alkalinity of the soil. For example, we can compare a soil's pH level with your stomach. If your stomach is too acidic it will not work well. This will then cause problems for you stomach and the rest of your body. The same is true with soil. In good conditions, the soil's pH level will be neutral, this will greatly improve the productivity of everything that is being grown in that soil.

pH Chart

If the soil is acidic, nutrients will easy leach out of the soil. Productivity will reduce and if the soil is very acidic, only a few types of plants can be grown.

If the soil is alkaline, there are many nutrients in the soil, but they are bound and not easily available for plants to use. Productivity will reduce and only a few types of plant can be grown.

By adding enough mulch, compost and other organic materials, the soil will contain more humus which will then neutralize the soils pH levels, as well as increasing the amount of nutrients in the soil.

Using chemical fertilizers when the soil is in acidic or alkaline condition will only be wasting money, because a lot of nutrients will be bound in the soil or leach out of the soil. Besides that, it will also create many more problems in the future.

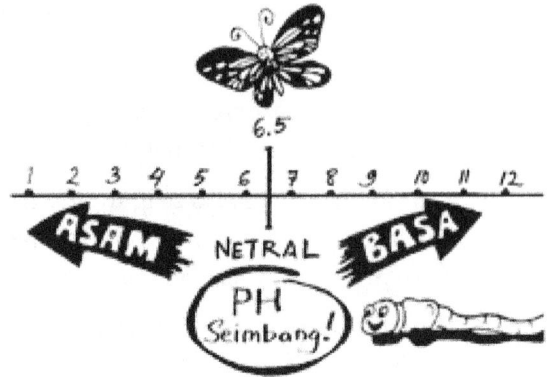

Identification of Soil pH

Acid soils:

⊙ Are generally found in wetland, areas with higher rain fall, and in the mountains

⊙ Taste sour, like vinegar

Alkaline soils:

⊙ Are generally found in dry land, coastal areas, and areas with lots of limestone

⊙ Taste sweet

Testing Soil pH

Soil pH can be accurately identified using a pH tester. A pH tester shows a series of numbers, ranging from 1 to 12. Number 1 shows that the soil is most acidic, and number 12 shows that the soil is most alkaline. The ideal soil condition will have a pH of 6.5 or neutral; in this condition, the soil is neither acidic nor alkaline.

There are a few types of pH testers. Some agriculture workers and NGOs may have this type of tester. However, by identifying landforms (for example, swamps), rocks, and common tree types, you can identify the soil pH without needing this equipment.

Solutions for Balancing Soil pH

The best solution for acidic or alkaline soils is to increase the amount of humus in the soil. This can be done by regularly using mulch, compost, liquid fertilizer and other organic materials. Increasing the humus content in the soil will make the soil pH neutral, allowing more nutrients to stay in the soil and be available for plant use.

Other Solutions for Acidic Soils

- Ash from wood fires (there must be no plastic content in the ash) can be spread over soil that is acidic. Don't use more than 1 kg for every 30 square meters each year. Don't burn grass and plant materials to make the ash; grass and plant materials are also very important for balancing soil pH

- For acidic soil in small areas, crushed seashells will provide lime to help balance soil pH

- For larger areas, dolomite can be used. Lime can also be used, but dolomite is better because it contains magnesium, and is safer for plant roots. These materials are expensive, and should only be used after the soil pH has been tested

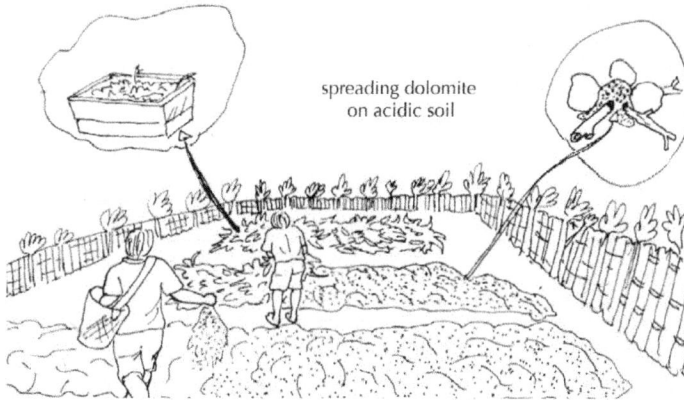

spreading dolomite on acidic soil

Amount (kg) of dolomite needed to raise soil pH levels to 6.5, per 30 square meters:

Soil pH	Sandy Soil	Loam Soil	Clay Soil
6.0	1 kg	1.5 kg	2 kg
5.5	2 kg	3 kg	4 kg
5.0	3 kg	4 kg	6.5 kg
4.5	3.5 kg	6.5 kg	9 kg
4.0	4 kg	8 kg	10.5 kg

balancing soil pH with green manure crops and pond soil

Other Solutions for Alkaline Soils

- ⦿ Use 6 kg of compost per square meter to lower soil pH by 1 point (for example, 8.5 pH to 7.5 pH). This does not need to be applied all at one time

- ⦿ Use 2 kg of manure per square meter to lower soil pH by 1 point

- ⦿ Iron sulfate ($FeSO4$) or other materials that contain sulfur can be used, but they are expensive. It is best to test the soil pH before using these materials

Amount (kg) of iron sulfate, or other materials that contain sulfur, needed to lower soil pH by one point per square meter:

Material	Sandy/ Loam Soil	Clay soil
Iron sulfate	2 kg	8kg
Material that contains sulfur	300g (1/3kg)	1 kg

4.5 NUTRIENT CYCLES

All plants need nutrients to grow. Some of these nutrients are stored in the plant's leaves, fruit, stems, trunk and roots as the plant grows. Trees and deep rooted plants are able to soak up minerals from deep down in the soil through their roots, but these minerals are sometimes not available in the soil. Trees will also soak up water from deep in the soil, like a big water pump.

Some nutrients are used for the process of the plant's growth, others are used by the plant to form fruits or seeds, or are stored within these fruits or seeds. This is the same for vegetables and other smaller plants.

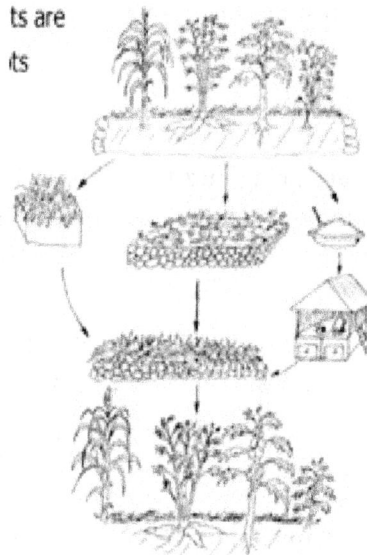

ts are

ts

These nutrients can become lost from the system (the soil), and need to constantly be replaced. A lot of nutrients can be recycled back into the soil through humans, animals, compost and mulch. Some nutrients that do become lost can be replaced by using some soil improvement techniques, such as:

- Planting seasonal and perennial legume trees
- Implement crop rotation and allowing the land lie fallow (not planting for a period of time)
- Using compost or liquid compost
- Using seaweed, manure, animal bones and carcasses, and other organic materials
- Applying mulch regularly
- Implementing a variety of systems, for example planting many trees, which besides functioning as a wind break will also attract birds and other wild animals, which will then naturally give manure to the land. You can also keep pigeons, their manure is easier to collect. Bird manure contains high concentrations of nutrients and is a very beneficial high quality natural fertilizer when dry.

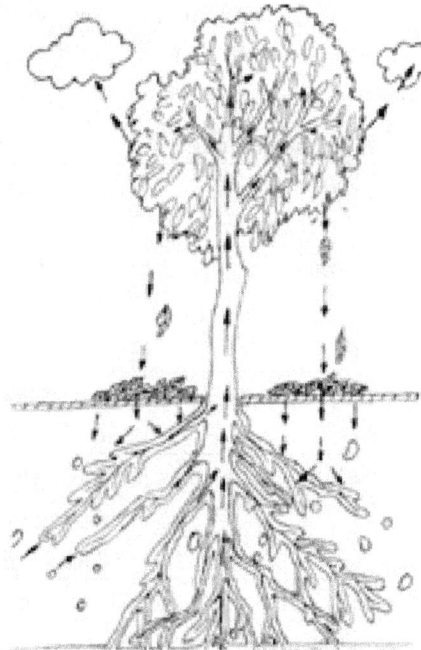

In tropical climates a lot of nutrients are stored in trees, and only a small amount are stored in the soil. Therefore, cutting down forest means removing nutrients from this system. The soil will only last for one or two years, after which it becomes poor in nutrients and not very good for growing crops.

In Indonesia, the amount of forest is continually reducing, mainly caused by clear cutting and burning. Forests are being cut down primarily for commercial purposes, and forest burning happens almost every dry season. Besides this, forests are being used for many other needs, such as agricultural land, new residential settlements, farm land, animal grazing land, and a source of fire wood.

Burning land is a very serious problem, because it reduces soil fertility and removes valuable nutrients from the soil. Each time the land, leaves, grass and other plant materials are burned, nutrients which are stored in plants become lost. This occurs both on agricultural land as well as animal grazing land. After burning, the ash does provide a small amount of potassium and minerals, but the nutrients that have been removed are much more than what is contained in this ash. To get potassium, using ash just from kitchen cooking fires is enough.

Remember, the more nutrients that are recycled back into the system, the less outside inputs are needed!

4.6 NUTRIENT DEFICIENCIES

In some places, a lot of soil is nutrient deficient. Some areas are very deficient, and others only lack one or two types of nutrients. Just like people, plants also need a range of vitamins and minerals to grow well. If nutrients are not available, plants will be smaller, and more susceptible to drought, pests and disease. Plants show specific signs when they are missing a nutrients, for example:

Missing nutrient	Plant Characteristics
Nitrogen	All leaves and new growths are yellow and pale
	Early maturing, fruit and flower size is smaller
Potassium	Leaves are small and darker in color
	Older leaves are blue / purple with yellow edges
	Plant growth is slow
Phosphorus	Fruit size is small and poorly colored
	Burnt leaf edges and yellowing of older leaves
Magnesium	Yellowing leaf edges, yellow spots but the leaf veins stay green
	Often there are brown spots on the leaves
	Old leaves drop of faster
Sulphur	All leaves have a dull color
Calcium	New leaves and shoots grow and develop poorly
	Unusual fruit growth
Micro nutrients	Symptoms vary

If plants are sick or not producing well, it is not enough to just add a basic fertilizer. In fact, this approach can even cause more problems. It is better to first try and identify the problem, and then to figure out what the exact deficiency is. In this way, problem solving will be more effective and inexpensive.

4.7 ORGANIC SOIL IMPROVEMENT STRATEGIES

If land is under cultivation, then nutrients are being used and must be replaced. To improve the nutrient condition, it is not enough to just replace the missing nutrients, but also with time there should be soil texture improvement so that the soil can store more nutrients and water.

Natural organic fertilizers can be used regularly and can be applied before, during and after planting. The nutrients that are not used will be stored in the soil to be used later. Both for short term and long term, organic fertilizers will help to improve the soil's condition.

It is always better to compost manures before using them as fertilizers. If the manure is fresh, especially bird manure, it can burn plants, especially small plants and young vegetables.

The nutrients are also not yet available for the plant to use. This is the same as humans trying to eat rice, corn or meat before it is cooked! Composting organic materials will concentrate the nutrients, making them easily available for the plant to use.

There are many different composting methods, some of which will be explained later on. Over time, by experimenting, you will find out what works best for your land, climate and needs. This could be new techniques, traditional techniques, or a combination of both.

4.8 NATURAL NUTRIENT SOURCES

Almost all nutrient deficiencies can be handled by using compost, liquid compost, and mulch. This is the best and most balanced method.

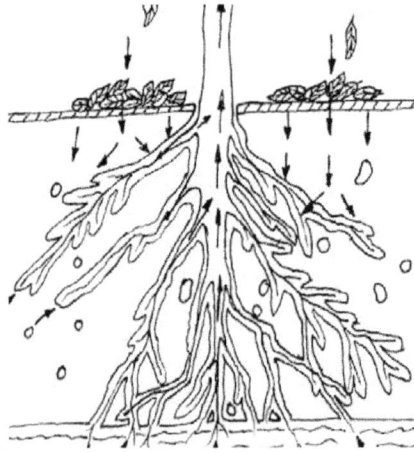

Sometimes, a specific nutrient or nutrients are not available in the soil or plants, because of erosion, deforestation or poor soil. This nutrient needs to be reintroduced into the soil, and for best results, added again once or twice a year. This new nutrient should be recycled within the system as much as possible to reduce the need for more outside inputs.

Some sources of new nutrients:

- ◉ Seaweed contains many nutrients. Many nutrients are washed out to sea with water because of land erosion or landslides. Seaweed is very beneficial and contains many nutrients to help to replace missing nutrients

- ◉ Bat, pigeon, chicken and duck manure are concentrated manures. Bat and pigeon manure are the best, but all manure is good because it comes from organic sources

- ◉ Animal bones, carcasses and innards are a high concentrated source of nutrients and can provide a lot of micro nutrients. These materials must be composted first, or buried under new fruit trees

- ◉ Mulch or manure from other areas

- ◉ Legumes to add nitrogen to the soil

- ◉ Wood ash from kitchen cooking fires can supply potassium

- ◉ The soil from the bottom of a well managed fish pond contains lots of nutrients

- ◉ Mulch from water plants. Water plants are very good at taking and storing excess nutrients from water. Pond water also contains nutrients

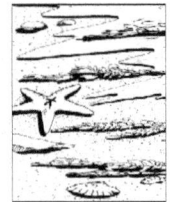

- ⊙ Tree leaves provide a variety of nutrients, because trees soak up minerals from deep in the soil

- ⊙ Micro-nutrient fertilizers (best if made from seaweed or rock dust) can be used to replace nutrients. This is not a normal fertilizer, such as urea fertilizer which does not replace certain lost nutrients

4.9 EM (EFFECTIVE MICROORGANISMS)

All organic composts provide microorganisms, bacteria, soil biota and fungus. All these components or elements are important for improving soil structure and quality. EM is a liquid that can be added to compost, liquid compost or directly into the soil. EM contains microorganisms which the soil needs.

4.9.1 EM is used to:

- ⊙ Speed up the composting process

- ⊙ Improve the quality of natural fertilizers

- ⊙ Make nutrients more available to the plants

- ⊙ Improve all aspects of soil quality

EM can be bought from agricultural stores. The bottle of EM that you buy can then be used to make more EM, because bacteria and microorganisms easily multiply. Only one bottle of EM is needed to have a continuous supply.

4.9.2 How to Multiply EM

Materials: Used water / aqua bottles, water, palm sugar, 1 bottle of EM

Method:

- ⊙ Fill the empty bottles with water. Add a slice of palm sugar and shake well, until the sugar dissolves. Add one full capful of EM

- ⊙ Gently mix and stir in 2. Leave this mixture in a dark place for 2 weeks. Avoid direct sunlight

Microorganisms and bacteria will multiply quickly because they will feed on the sugar. This new EM is now ready to be used, and can be used to make new bottles of EM.

4.9.3 How to Use EM

Liquid compost: Add about 1 bottle of EM to one drum of liquid compost.

Compost: Add about 1 bottle of EM to a small amount of compost or 2 bottles of EM to larger amounts of compost.

Soil: Spray EM liquid to agricultural land and around plants. Only a little EM is needed because these microorganisms will multiply on their own. It is more effective to use EM at the same time as when mulch and compost is added.

Rice paddies: Add a few bottles of EM into the irrigation water. This will be much more effective if combined with SRI (System of Rice Intensification) techniques.

4.10 LIQUID FERTILIZER

Liquid fertilizer can be easily prepared and is very useful in many ways, including for nurseries, small gardens, fruit trees and other large crops. This is a good way to make nutrient rich fertilizer from small amounts of manure and other organic materials. Liquid fertilizer can easily be sprayed over large areas of land.

Liquid fertilizer is made in very strong concentration, so it needs to be mixed with water before being used. Liquid fertilizer can be stored and it lasts a long time, and can be used on larger areas of land. Liquid fertilizer can be made in a container of any size, from a bucket to a large drum. The more you make, the better. This fertilizer can be made from any organic material, and can be stored anywhere, as long as it is protected from hot sun and hard rain.

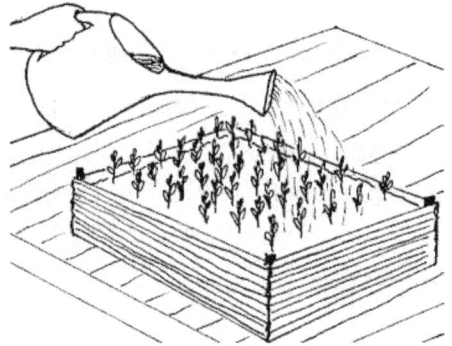

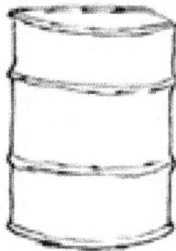

4.10.1 How to Make Liquid Fertilizer

⊙ Prepare a container, for example a drum. Make sure the drum does not leak, cut off the lid by cutting around its edges and then hit down any sharp areas along this edge.

⊙ Clean the inside of the drum using detergent, lemon, and water, then dry the drum in the sun for 2-3 days. Make sure that all oil, petrol or other poisonous materials are gone, because when the liquid fertilizer is made bacteria will live in it; poisonous materials can kill this bacteria.

⊙ Fill 1/3 of the drum with green grass (weeds), green leaves (legume cuttings), or seaweed (if you live near the ocean). Using weeds in liquid fertilizer will give multiple benefits, because besides the weeds providing many useful minerals, this also reduces weed problems.

If using legumes, don't put branches in the liquid fertilizer, because these branches take a long time to rot and make it difficult to stir the fertilizer. Seaweed contains nutrients and minerals that are useful and important for plants. Sometimes these nutrients are lacking from the soil, manure, and plants. Seaweed must be washed first to remove the salt, because salt can have a bad effect on soil quality and plant growth. When collecting seaweed, only collect the fresh seaweed, as dry seaweed contains much more salt.

⊙ Fill the next 1/3 of the drum with manure. Fresh animal manure contains more nutrients than dry manure.

Combine different types of animal manure (if available) to achieve the best result, as different manures contain different types of nutrients. Bird manure is best, and then pig, goat, cow and horse manure.

The smaller the animal, the stronger the manure (mouse manure is really great, if you can collect it). Therefore, less bird manure is needed than cow or horse manure.

⊙ Add 2-3 shovels of healthy, living soil. Healthy soil contains many biota, which will speed up the process of turning organic materials into fertilizer and help prepare the nutrients for plant use.

When soil biota / bacteria eats the organic materials in liquid fertilizer, this bacteria will continue to multiply. Putting bacteria into the soil is just as important as providing nutrients for plants.

⊙ 6: Fill the container with water.

⊙ 7: Other materials that can be added include: ½–1 shovel of kitchen cooking ash, to add minerals and potassium, and 1 shovel of fishpond soil.

If easily available, animal carcasses are also useful, like: Rat carcass, fish heads and bones, chicken carcasses, and smaller animal innards can all be added to the liquid fertilizer. This will add nutrients and minerals to the fertilizer. Remember, the most important thing is to provide more bacteria which will speed up the rotting of organic materials in the fertilizer.

9: Cover the drum to prevent animals, like mosquitoes and flies, from entering, to avoid direct sunlight which could kill bacteria, and to avoid rain entering the drum.

10: Stir the liquid fertilizer using a long stick for 5-10 minutes every day, for 2 weeks. This must be done to add oxygen to the fertilizer. In this fertilizer there are two types of bacteria, aerobic and non-aerobic. Aerobic bacteria needs oxygen, while non-aerobic bacteria does not need oxygen. Both bacteria work in the same way, but aerobic bacteria works better to create quality fertilizer, because its decomposing process is even and it reduces bad smells. So, the more often you stir, the faster the decomposing process and better the fertilizer quality.

4.10.2 Using Liquid Fertilizer

To use liquid fertilizer, it must first be mixed with water. Combine 1 part liquid fertilizer with 20 parts water (1: 20). If this fertilizer is not first diluted with water, and directly applied to plants, especially in large amount, it will burn the plants leaves and roots because the fertilizer concentration is still too strong. Young plants are generally more sensitive than older plants.

Use liquid fertilizer once or twice a week for vegetables and small trees, on other days it is enough to just give water. For fruit trees that are already established, use liquid fertilizer once or twice a month. When watering, you can use a container, like a can, with holes punched into the bottom.

For vegetables, first apply mulch around the plant, and then spray with fertilizer. If possible, avoid spraying the leaves directly, don't let the still concentrated liquid fertilizer burn the plant. This is also important with trees.

If the liquid fertilizer is almost finished, you can add more organic materials. Don't forget to continue stirring this liquid fertilizer, and wait two weeks before use. Don't wait until the fertilizer is completely finished, because then you will have to repeat the entire process from the beginning.

4.11 COMPOST

Compost is made up of decomposed organic matter, which is a concentrated rich nutrient source. The main ingredient is carbon and a small amount of nitrogen, as well as other nutrients, minerals and soil biota.

Compost doesn't just provide nutrients for vegetable and fruit plants, but it also improves the soil quality. There are many ways to make compost, from simple mixtures of rice husks and cow manure, to ones that are made from a variety of materials. Use available materials, and experiment for yourself.

4.11.1 Making Quick Compost Heaps

- A compost heap will work well if it is made all at once. This means all the materials are collected, and then the compost heap is made layer by layer

- A compost heap should be about 1 m wide, 1 m long, and 1 m high to create enough heat to decompose well

- You can use many different materials, such as: Animal manures, grass, weeds, water plants, leaves, seaweed, soil, rice and coffee husks, sawdust, animal carcasses (mice, fish, chickens and so on), urine, small bits of paper which can be broken down, and other natural materials which can be easily found

- Try to combine about 1/3 manure, 1/3 dry leaves, 1/3 coffee husks, rice husks or sawdust. These amounts do not have to be exact, just estimate

- Use a small amount of bird manure (pigeon, chicken or duck), or a larger amount of animal manure

- These materials should be in layers, 5-10 cm thick. The first layer of organic plant material (dry leaves, husks and other dry materials) is 10 cm thick. The second layer is of decomposed animal manure. The third layer is a little kitchen ash and healthy soil. Then add some water. Repeat these steps until the heap is 1 m high, 1 m wide, and 1 m long

⊙ Add water twice during the process of making compost. First, add about 2-3 buckets of water during the beginning process, than add 2-3 buckets more when the compost heap is already finished. Water is very important for the composting process, but not too much. The compost should be moist, but not wet

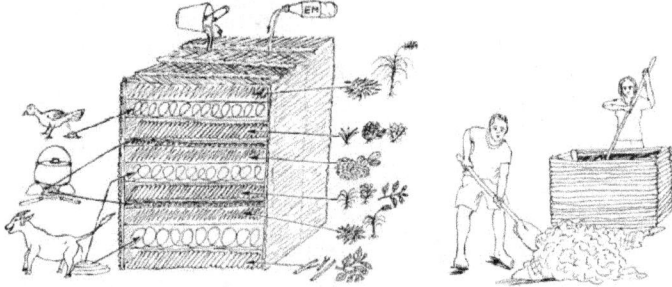

⊙ Cover the compost to protect it from rain and direct sunlight, and keep it somewhere safe from animals, this will also help to keep the heat in the compost. This cover can be made of palm fronds, an old tarpaulin, banana skins or leaves, or a thick layer of leaves. If possible, have this cover ready for the wet season to prevent rain from entering

⊙ If all these steps are followed, the compost will become very hot (around 65°-68° C), because of bacteria activity in decomposing. This temperature will cool down again to around 45° C after 2-3 weeks • The compost now needs to be turned inside out, because the outer layer is not receiving enough heat and will not be as composted as the inner layer. After the compost is turned, add 2 more buckets of water, but only if needed, then return the compost cover. The compost heap will then reheat and be ready for use in 2 weeks

⊙ Now the compost is ready to be used in the garden. Use 2 handfuls for every seedling, and 4 handfuls for every mature plant. It is even better to cover the entire garden area with a layer of compost, about 3-5 cm thick will work well

4.11.2 Making Slow Compost Heaps

Slow compost can be made from only 2 or 3 types of material, but you need the balance the amount of manure with plant materials. A good mixture is about 1/3 manure with 2/3 plant materials. Slow compost will take about 2-3 months before it is ready to be used. This type of compost is not as nutrient rich, but it is still good enough to improve soil condition.

4.11.3 How to Use Compost

⊙ Start to make a new compost heap when the old compost heap is already half used up. This will ensure a continuous supply of compost

⊙ Use compost on plots 2 weeks before planting vegetable seedlings or directly planting seeds, such as corn, beans, eggplant and so on, to allow the nutrients and compost soak into the soil. Wait 2 weeks after planting before adding more compost

⊙ The best place to use compost for trees is directly under the outer layer of leaves (on the ground of the tree's outer crown, not around the tree trunk). This place is around the bottom of the trees outside leaves, and it is where the tree's roots feed from most. Don't compost around the tree trunk, because this could cause the trunk to rot. This is true for all types of fertilizers

4.11.4 Making Compost Baskets and Trenches

⊙ Dig holes in the middle of the garden plot, about 1 hand length deep (15-20 cm), and 3 hand lengths wide, leave 1 meter in between each hole (about 1 large step). Otherwise, the trench can be dug lengthwise in the middle of the garden plot, about 1 hand length deep, and 2 hand lengths wide.

⊙ Place bamboo sticks around the edges of the holes or trench. These sticks should be about 1 hand length apart and 2 hand lengths above ground.

⊙ Make a fence of woven bamboo / leaves through the sticks.

⊙ Put into the holes or trench in order:

  • A thin layer (about 5 cm) of small branches or dry grass to provide air

  • Different types of manure

  • Grasses, weeds, leaves, and washed seaweed

  • Add 1 handful of kitchen ash to each hole, or per meter of the trench

⊙ As the old compost materials decompose, add new materials, in the same layers as before. The compost does not need to be turned.

⊙ Plant seeds when the compost at the very bottom begins to decompose. Leave about 1 hands length in between the seed and the compost basket or trench.

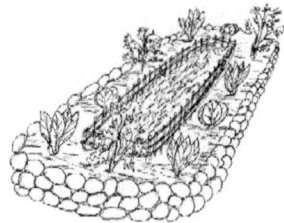

- Water the plants regularly for 2 weeks after planting. Then you can water the compost basket or trench directly, not the plants. The plant roots will grow into the hole or trench. This will improve plant growth and save water.

- When the plants are harvested, the compost from the baskets and trench can be shoveled out and used on the garden plots to add humus to the soil. After this, new compost materials can be added to the basket hole or trench in preparation for the next planting.

4.11.5 Compost Pits

A compost pit is a great way to supply plants with nutrients. Examples of plants that work well with this system are bananas and papayas. Materials for this compost uses anything available, such as: Leaves, weeds, manure, rice / coffee husks, and paper can all be added. Urine is also recommended. The compost that collects at the bottom of the pit can be dug up each year and used for plants.

Benefits of using compost pits:

Benefits of using compost pits:

- Deals with weed problems

- Uses excess water and organic wastes

- Stores more water in the soil and in the compost materials, so that less water use is needed for plants, especially in the dry season.

To deal with mosquito and insect problems inside the pit, soak a handful of neem leaves in a bucket of water for 2 days, then pour this liquid with the leaves into the compost pit. Repeat this method every 3 months.

Direct Composting

Compost can also be placed directly on top of the garden plots or land where the garden plots will be made. The soil underneath will receive the benefits of nutrients that are released, and this will increase the amount of soil biota.

4.11.6 Diluted Urine

Human urine is an easily available, free and continuous source of nutrients. Urine contains quite high amounts of nitrogen. If urine is diluted with water (10-20% urine, with 80-90% water), it becomes a great fertilizer for fruit trees, citrus trees respond especially well. Urine can also be added to compost pits and other types of compost heaps.

Urine is not recommended for use in vegetable gardens. Also, mature fruit trees will receive benefits from direct urination, but not continuously on the same place.

4.11.7 Earthworm Farms

More worms in the soil means better soil. Farming earthworms is a simple way to quickly increase the number of earthworms in your soil.

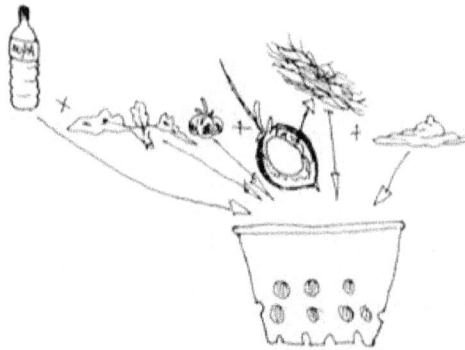

4.11.7.1 Materials needed:

- ⦿ One old bucket
- ⦿ Coconut husk fiber
- ⦿ Cow or horse manure that has been soaked in water
- ⦿ Left over kitchen vegetables
- ⦿ A flat piece of wood or metal
- ⦿ A large rock

4.11.7.2 How to Make an Earthworm Farm

- ⦿ Make 10 coin sized holes at the bottom and on the sides of the bucket
- ⦿ Dig a hole in the garden, large enough for the bucket to fit inside of. The top of the bucket should be about 1 hands length above soil
- ⦿ Fill the bucket with coconut husk fiber, left over vegetables, and lastly with animal manure. Use about the same amount of each material

⊙ Cover the bucket with a piece of wood or metal so that animals cannot enter, and then place the large rock on top to hold it in place

Make sure that the materials in the bucket are moist, especially during the dry season. Add more materials to the bucket if needed. Every few months, clean out the bucket, the materials in the bucket can be used as they have become a rich nutrient compost. After, add new materials to the bucket. Earthworms will come and eat the materials in the bucket, and then return to the soil. Earthworms are useful for small and large gardens, and even for rice paddies. With more earthworms, the soil becomes better and plants will be healthier!

4.11.8 Mulch

In natural forests, leaves, rotting materials, animal manure and even dead animals, all make up a mulch which covers the ground, like a skin. This skin is continuously being added to and also is continuously decomposing. Mulch (or skin) provides nutrients and humus to the soil as it decomposes, which are then used by plants and trees as food. Besides that, it also continuously provides food for soil biota.

We can copy nature by using mulch to make a skin for the soil. This skin is an important natural protection against drying from the sun and erosion because of rain. This skin also provides food for the soil biota in your garden.

Mulch that is used on gardens, agricultural land, crop land and reforestation areas can be grass cuttings, tree prunings, leaves, compost, decomposed manure, rice / coffee husks, used paper, rocks, animal bones, or any material, so long as it is organic. Make sure there is no plastic rubbish, used batteries, glass bottles, or any other non-organic materials.

4.11.8.1 Benefits of Using Mulch

- ◉ Keeps soil temperatures stable, which means that the soil temperature is cooler in hot temperatures and is warmer in cool temperatures. This moderate temperature is good for plant growth. Remember, this is the same as with people!

- ◉ Reduces weeds. Weeds can only grow if there is light, so without light the seeds of weed plants will die. A layer of mulch will prevent sunlight from entering

- ◉ Provides organic matter, and valuable nutrients for the soil

- ◉ Mulch will become humus, which will improve soil structure and increase the number of soil biota

- ◉ Increases the soil's water storage capacity

- ◉ Helps to neutralize the soil's pH levels

- ◉ The soil will become easier to dig and manage

- ◉ Reduces erosion

- ◉ And of course, all of the points above will help to increase production!

4.11.8.2 How to Use Mulch

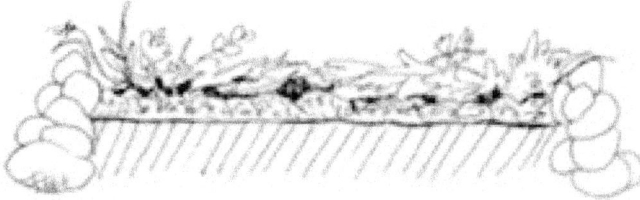

- ◉ Apply mulch regularly and as thick as possible. 5- 10 cm is generally the ideal thickness, but for fruit trees up to 20 cm thick is better

- ◉ Apply mulch to the soil before planting

- ◉ Apply mulch to the whole plot, not just around the vegetables and plants

- ◉ Use fine textured mulch for vegetable plots, and a coarser textured mulch for mature plants and trees

- ◉ For trees, apply mulch underneath the outside leaves, because this is where the trees roots will feed. Regularly applying mulch will improve the tree's health, and the size and amount of fruits

- ◉ Don't let the mulch touch the plants stem or trunk. This is very important to avoid rotting, especially in the wet season

- ⊙ Use rocks, branches, or whatever material th This will help hold the mulch in place and p

- ⊙ If you are using mulch and compost at the same time, apply the compost underneath the mulch to maximize the benefits of the compost

- ⊙ If you are using weeds as mulch, separate the weed seeds first and give them to animals or use in liquid compost. This will reduce future weed growth

- ⊙ Plant plants that can be used for mulch, like legumes. Remember to always think of the most multifunctional plants, for example, plants that can produce mulch material but also provide food for humans or animal fodder, function as a windbreak, fence, or help to reduce erosion, improve soil, produce fire wood, building materials, and so on

- ⊙ Leftover farming materials, like rice / coffee husks, can be composted or dried before being used as mulch. These materials should be put in a pile for 1 month or more before being combined with manure to make compost, then use this compost as mulch.

4.12 LEGUMES

- ⊙ Legumes are a type of plant that gives nitrogen to the soil. There are many different types of legumes, some are annuals and others are perennials. This plant is a very important part of any land or system, and can be used in many different ways.

- ⊙ Legume plants bind nitrogen from the air in soil to nodules, which are attached to the plant roots. These nodules are very small, about the size of a match head or smaller. The nodules provide nitrogen for the legume plant. Excess nitrogen which the plant cannot use is let out into the soil, and is available for other plants to use.

- ⊙ Bacteria called rhizobium attach themselves to the roots of the legume plant and live there. This bacterium is only released into the soil after the roots die.

4.12.1 Types of Legume Plants

- ⊙ Annual legumes: All beans, all peas, clovers.

- ⊙ **Perennial legumes:** All types of acacias, leuceana, casurina, sesbania, moringa, gliricidia, tamarind. Legumes provide many benefits. Some legume products include: food, animal fodder, mulch and compost material, timber, fire wood, and medicines. While legumes can also function as: windbreaks, living fences, trees for shade, and trellises. Legumes can be planted together with other plants / crops.Prune 3 or 4 times a year. If a legume trees branches are pruned, the roots will also die back to the same amount that is pruned. Therefore, the dead roots with nodules will release all of their nitrogen into the soil. Other plants can then use the nitrogen that is released. The prunings can also be used as mulch, animal fodder or compost material. As the legume grows back, its roots will also grow back, and new nodules will grow on them. This is a sustainable cycle. If a legume tree dies, it will still provide nitrogen from its roots for a few years afterwards.

- ⊙ **Annual legumes:** Prune back after the first flowers grow, this is because these plants need a lot of nitrogen when producing seeds, so there won't be much nitrogen left in the soil for other plants.

4.13 TECHNIQUES FOR USING ANNUAL LEGUMES

4.13.1 Crop Rotation

Different crops use different amounts of nutrients to grow. If you grow the same type of crop over and over again on the same plot of land, some nutrients will become depleted. The soil and its nutrients will then become imbalanced.

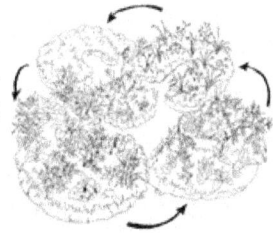

Some examples of plants and the nutrients they use to grow:

- ⊙ Pumpkins and melons like as much nutrients as they can get
- ⊙ Corn and tomatoes use a lot of nitrogen and some other nutrients
- ⊙ Vegetables use less nutrients
- ⊙ Beans and peas use nitrogen and other nutrients, but they also produce nitrogen
- ⊙ Carrots and radishes use less nutrients

It is good to rotate plants from plot to plot each season, or even better, to grow different types of crops together. Crop rotation will also help to reduce pest and disease problems. It is important to let the land lie fallow during a crop rotation cycle. Fallow time or 'rest time' means that nothing is grown for a period of time. During this fallow time, mulch, manure and compost can be applied and left on top of the soil or turned into the soil. Animals can also be used during a fallow period to add manure, improve the soil, and reduce weed problems. Fallow time allows the soil to 'recover'.

4.13.2 Green Manure Crops

During the fallow period, the soil can be improve(
be planted by planting green manure crops. This gr
thickly and should cover the whole land area. Gr
legumes, annual grasses and other annual plants. G
food for humans, but only for the soil's benefit. As
down the green manure crop and leave it on the lan(
time and labour, and maybe even a little bit of money

- Nitrogen from legume roots is left in the so
  planted

- Mulch and compost material

- An increase in humus and soil biota

growing and cutting green manure crop

These benefits all help improve soil structure and
crop. So, the increased productivity of the next crop]
have spent on the green manure crop.

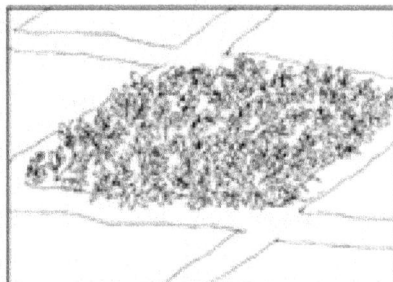

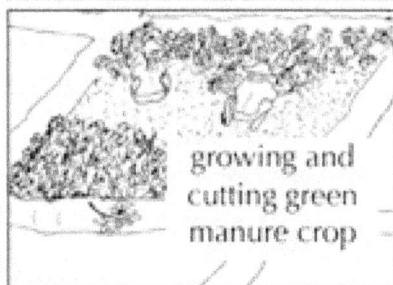

4.13.3 Annual Crop Integration

Legumes can also be integrated with vegetable crops or other crops at the same time. This technique will increase diversity in crops harvested from one plot of land. The other crops growing will benefit from the excess nitrogen which is produced by beans and peas. Crop integration can follow any pattern, as long as it suits the plants and shape of the land.

4.14 TECHNIQUES FOR USING PERENNIAL LEGUMES

4.14.1 Living Fences

Legumes are easily grown from seeds or cuttings, if the plant is continuously pruned it will become a thick living fence. Living fences can also function as windbreaks for vegetable plots, chicken and animal systems, aquaculture, and nurseries. This plant grows quickly and the material from pruning is very useful as animal fodder and material for mulch and compost. Living fences will also protect crops from animals.

4.14.2 Legume Swales and Terraces

On sloped land, legumes can be used to prevent erosion. Plant legumes to create living swales or terraces:

- Plant the seeds following the contour of the land

- Plant them very close together (3-5 cm)

For more information about tree swales and terraces, see Module 8 – Forests, Tree Crops and Bamboo.

4.14.3 Perennial Crop Integration

Legume trees can be integrated with fruit trees, coconut trees, coffee plants, vegetables and other crops. Legumes will provide nitrogen, mulch, as well as protection from winds and erosion. All of these plants can be integrated in rows, plots, or combined in other ways. It is all up to you. Taller legume trees, like Casurinas, provide shade for coffee, vanilla, and other plants that like shade. Animals also need shade, and legume trees can provide this function.

4.14.4 Pioneer Trees

Pioneer trees are trees that are planted first in a system. They are used to prevent erosion, provide protection for future crops, improve the soil and provide mulch. Pioneer trees often grow in hard soil conditions. Because of this, use trees that don't need a lot of water, and are most endurable in hard conditions. Legume trees can be pioneer trees, because they:

- Easily grow from seeds or cuttings

- Grow quickly

- Provide nitrogen for other crops.

4.15 FERTILE SOIL

If we have fertile, healthy soil, this will be a strong foundation for continually producing productive sustainable crops, whether for a small home garden or a large agricultural farm. If the soil is well managed in a healthy farm system, the soil will become more fertile over time and continually produce healthy food and crops.

- Almost all types of legume trees have short life spans. Use short life legume trees as pioneer trees, and longer life legume trees for windbreaks, shade and living fences

- Plant many more legume trees as pioneer trees, and then after a year or two, cut down the weaker or smaller trees to be used as mulch and to provide space for new trees

- If legume trees are planted on sloped land, when their nitrogen is released into the soil, it will slowly move down to lower areas, so lower areas will receive enough nutrient supply

- Remember, legumes are multifunctional. They provide many benefits. Use as many of these benefits as possible

4.16 NON-ORGANIC FERTILIZERS

In the last few decades, the use of non-organic fertilizers has increased dramatically all over the world. In Indonesia, the practice of using non-organic fertilizer is supported by a government program to increase farming produce. So, almost all Village Cooperation's (KUDs) and some farming stores sell non-organic fertilizers directly to farmers. However, besides the high prices of non-organic fertilizers, even if used correctly by farmers, they only provide short term benefits, which will not last in the long run.

The information above is only the writers opinion, and an attempt to analyse and suggest to farmers that if we do use chemical materials or non-organic fertilizers, we should truly understand the negative effects of the materials on our environment. Farmers in Indonesia often have difficulty accessing information about these topics, in part because there are not a lot of field workers that really understand such topics.

Once we damage the land, it will take a long time for the land to recover to its normal condition, when of course food production for community needs must still be met. If non-organic fertilizers are being used, water irrigation must be good enough to supply even more water for crops, because non-organic fertilizers stimulate plants to soak up more nutrients and water from the soil than what they normally need for growth. Non-organic fertilizers are similar to stimulating medicines.

If the practice of using non-organic fertilizers continues, in time the nutrients in the soil will be used up, the amount of acidity in soils will increase, bacteria / microorganisms in the soil will die, soil structure will deteriorate, and in the end farmers and the soil will be dependent on nonorganic fertilizers indefinitely. Also, many non-organic fertilizers do not provide all the nutrients and minerals the plants need to grow.

Even in highly developed countries, where there is good access to information, many problems are still experienced due to use of non-organic materials, let alone in still developing countries.

Some serious problems that can happen are health problems (sometimes skin, lungs, and even cancer), and problems to do with soil, such as too many nutrients in the soil. If there are too many nutrients in the soil:

- ⊙ Nutrient 'lock up' will occur, which means that there are nutrients stored in the soil, but they are not available to plants for use

- ⊙ Excess nutrients will be wasted, especially nitrogen, and washed away by water into wells, ponds, rivers and oceans. This also causes problems for water quality, which can affect fish and water plants, animals, humans and all living things in the surrounding environment

Another thing to consider, besides the negative impacts on the environment and ourselves, is that non-organic fertilizers are expensive, while it is not definite that a farmer's produce will sell to the market. This can make it difficult for farmer to pay back the money or loan they have used to buy non-organic fertilizers.

5

SEED SAVING & NURSERIES

5.1 INTRODUCTION: AN OVERVIEW

Saving and using local seeds is one of the most important methods for strengthening agriculture and increasing plant diversity.

Why? Because:

- It is inexpensive and easy to do, anyone can collect and save seeds
- It will increase the amount and range of foods that can be grown
- Seeds are valuable, they can be exchanged with other seeds or sold through a community seed bank
- Plant quality will naturally improve from year to year

Local seeds are adapted to local conditions. As these seeds grow in the local climate and soils, they become stronger. For example, if someone from Indonesia goes to live in England, it will take a long time for him / her to adapt to the cold climate, the people, language and culture.

It is the same with seeds and plants. The plants which grow healthiest and strongest are the plants that can become the seed source.

5.2 POLLINATION

- Pollination is a process a plant uses to produce fruits and seeds. During pollination, the pollen from the male part of a plant fertilizes the female part of a plant. This pollination process usually happens in a plants flower. Once the female part of plant is pollinated, the plant will produce fruit and / or seed. Different pollination techniques are used by different types of plants, for example:
- Beans, lettuce, tomato, cabbage and chilli are all plants which have male and female parts inside the same flower

⊙ Pumpkin, melon, cucumber and corn are plants which have male and female parts separate, on the same plant. These plants need insecs, wind, or human hands to fertilize

⊙ Papaya and salak are plants which have separate male and female plants, these plants need more than one plant to fertilize.

⊙ The more insects there are in the garden, the more pollination will happen. Flowers, food, plant diversity and ponds will attract insects to your garden.

Cross Pollination

Cross pollination is pollination that happens between plants, where the pollen from the male part of a plant pollinates the female flower of another plant. This process can happen naturally or it can be induced. Cross pollination of different types of plants happens when two types of closely related plants pollinate each other, for example:
Two different types of green leaf vegetables, two different types of corn, or a pumpkin plant and a squash plant. If this happens, the seed that is produced may be good, but more often the seed will be weaker than the parent plants, or may not grow at all when planted. Therefore, it is best to avoid cross pollination happening.

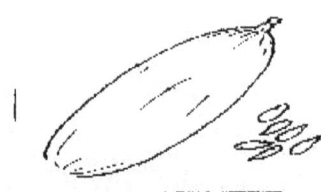

A few techniques for reducing the chance of different types of plants cross pollinating:

⊙ Plant one type of crop each planting season. For example, one type of corn or one type of eggplant

⊙ Green leaf vegetables, lettuce and cabbage flower at the end of their life. Let only one type of green vegetable, one type of lettuce, or one type of cabbage reach the stage to flower and produce seeds

⊙ If different types of plants are further apart, and many other plants are planted in between them, the chances of cross pollination are reduced

⊙ Hand pollination, such as of pumpkins, melons, luffa and cucumber will allow you to choose the type of seed which will be produced

You can use induced cross pollination to try and create a new type of plant. But be careful, the results could be something unknown, which could either grow well or not at all.

Some types of plants have both male flowers and female flowers. The female flowers have a small fruit below them.

In the afternoon, choose a male and female flower that are just about to open. Tie them, so that insects or bees cannot enter.

The next morning, open the flowers. Carefully pick the male flower, pull off the petals, and rub the pollen covered middle (stamen) inside the female flower. Again, tie the female flower.

When the fruit starts forming, tie a piece of string or material around the base of the fruit as identification.

Repeat this process on other female flowers, but use male flowers from different plants of the same type, to help keep seed quality and diversity.

5.3 INTRODUCING NEW VARIETIES OF SEEDS AND PLANTS

5.3.1 INTRODUCTION

New varieties of plants can be grown to add diversity of plants. Sometimes introducing new types of plants, grains, fruits and vegetables can increase crop yields. Don't forget to label every new type, so that it can be easily identified.

If a new type of seed or plant is introduced:

◉ Always use non-hybrid varieties of seed. Non-hybrid seeds can be saved and planted again each year, but hybrid seeds must be bought every planting season. Hybrid seeds are produced from two or more varieties of plants. Hybrid plants do not produce seeds that can be replanted. If they do produce seeds, the seeds will not be the same type of plant and the quality of the next crop will be poorer

◉ Plant test crops first to find out if plants grow well and produce well. A simple test is to plant 3 small plots of the new type of vegetable in the garden, each plot should be 3 m x 1 m. Each test plot should be in a different location, but grown using the same techniques. If the plants grow well, they can be planted in larger plots next time. This will help increase crop variety, but will still save a lot of time, work and money if the crop does not grow well

- ⦿ **From one area to another.** Check the seeds for insects or insect eggs. Remove seed pods and any plant materials. Wash the seeds until clean and dry. Cover the seeds with wood ash to avoid insect problems. A small amount of dried neem leaf, crushed and mixed in with the seeds, will help to kill insects and their eggs

- ⦿ **From overseas.** A countries government should give more attention to the quarantine department. The regulations should be followed to ensure that new diseases and pests do not enter

5.3.1 Potential Problems

New plants that are introduced are at risk of becoming a problem in the future, for example if the plant spreads quickly and becomes a weed which could disrupt the local environment. This can happen with any type of plant, even plants that are very productive can become a problem.

5.3.2 Research the following about any new plant before it is introduced:

- ⦿ What is its growing habit?

- ⦿ Does it spread naturally? For example, do animals spread the seeds?

- ⦿ Has the plant caused problems in other places?

- ⦿ Does the plant suffer from diseases which could spread to new areas?

This is very important for protecting our environment and resources for the future.

5.3.3 Seed Saving

By saving and storing seeds well, you will have disease free, good quality seeds that can be planted from season to season.

5.4 PRODUCING GOOD SEEDS:

To produce good seeds follow these steps:

5.4.1 Healthy Plants

To produce quality seeds, the first step is to grow healthy plants. To do this you will need healthy soil, and enough compost and mulch.

5.4.2 Choose the Best Seed

Always collect seeds from the best plants.

These plants usually:

- ⦿ Produce healthy and tasty fruits or leaves

- ◉ Are disease free and naturally pest resistant

- ◉ Are able to survive in extreme conditions. For example, are able to handle very dry or hot conditions, or can still grow well in rocky soils

Are slow at producing seeds. Collect seeds from plants that produce flowers and seeds last, not first Select seeds from many plants. If you are growing trees, for example teak trees, collect seeds from many different teak trees. It is the same with all other plants. When collecting seeds, remember that you will pass on the plants characteristics to the next crop. If you choose healthy plants, the next crop will have the same characteristics as the parent plants. Larger seeds will generally last longer than small seeds.

5.4.3 How to Harvest Seeds

Label the plants that seeds will be collected from, so that these plants won't be harvested for food. Wait until the plants are ripe to pick the seeds. This means leaving the plant until it is past the edible stage. Young fruits have young seeds, which may not germinate. The best time to pick seeds is mid-morning, on a clear and sunny day. If harvesting seeds in the wet season, you can pick the fruit, seed, or even better the whole plant, and hang it to dry near a fire. Even a small amount of moisture can damage seeds.

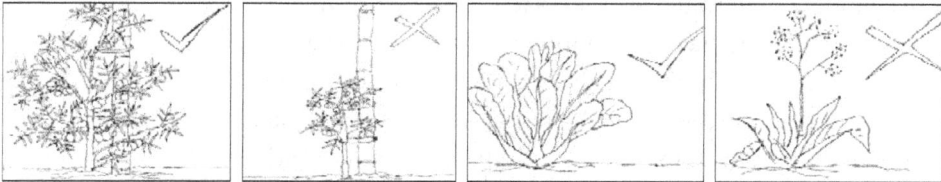

Plant	When to harvest seeds	How to harvest seeds
Tomato, eggplant	When ripe on the plant, slightly soft but not rotten	Hand pick the best fruits from the best plants
Cucumber, melon	One month after you would pick for eating (so the seeds are mature)	Hand pick the best fruits from the best plants
Capsicum, large chili	When ripe on the plant, or when red	Hand pick the best fruits from the best plants
Lettuce, green leaf vegetables	Wait until the seedpods are brown and dry, but not yet open	Cover in bag, then cut the main stem, so that no seeds will fall during collection
Beans, corn, sunflower	Leave the seeds to dry on plant during the dry season, pick when ripe in wet season and dry near a fire	Hand pick when the seeds are ready
Pumpkin	When ripe on the plant, wait 2-3 weeks before removing seeds	Hand pick the best fruits from the best plants

5.4.4 Cleaning Seeds

Separate seeds that have a dry pod or shell and remove them by hand. Small seeds with a shell can be kept in a bag, which can be gently rolled and carefully crushed to separate the seeds. Separate any plant materials from the seeds by winnowing or by hand.

Tomato, cucumber and pumpkin seeds can be removed and placed in a container with water. The seeds must be cleaned well and rinsed, so that all the plant flesh is removed from the seed. The seeds can then be dried.

Tomato and cucumber seeds can be fermented to remove some diseases. Remove some seeds and flesh from a ripe fruit. Place in container with water, leave for a few days. Foam will form on the surface showing that fermentation has happened. The seeds can then be washed with water. All remaining fruit flesh should be removed. Spread the seeds onto a plastic, wood or metal plate, and put them in the shade to dry out with the help of the wind.

5.4.5 Drying Seeds

Drying seeds is a very important part of the seed saving process. If the seeds are not dried properly, they will go rotten when stored. The seeds can be dried anyway you choose. However, to achieve the best results, it is very important that you follow these practical guidelines:

- Spread the seeds and air out. Shallow bowls, woven trays, old paper, woven mats or any other container can be used to hold the seeds. For larger seeds, place in woven bags and hang to dry. Turn them once or twice a day so that all the seeds can dry

- Protect the seeds from animals, especially mice

- For small and light seeds, give extra protection from the wind because they can easily be blow away

- Small seeds generally need about 1 week to dry properly, and larger seeds need about 1-2 weeks to dry properly

- Start the drying process for two days in shade or inside. After, move the seeds out into the sun for half of each day. This will help to kill insects and their eggs. Move the seeds inside at night. In the wet season, it is better to dry seeds near a fire Use a bite test to check if the seeds are dry or not. Bite a seed slowly. If the seed is hard and does not have a bite mark, than it is ready to be stored. If there is a bite mark, then the seed is not yet completely dry and needs to be dried for longer. If you tooth breaks when biting the seed, then next time you bite test seeds, don't bite so hard!

5.4.6 Storing Seeds

After the seeds are dry, they need to be stored well. If the climate is not ideal, seeds may easily rot if not stored correctly.

When in storage, seeds must be protected from:

- Air, which reduces the seeds lifetime

- Moisture, which can make seeds rot

- Heat, which can reduce the number of seeds that will grow when planted

- Animals, which can damage seeds

- Insects, which can eat or damage seeds. If insect eggs are laid inside the seed storage container, they will hatch and young insects will eat the seeds

- Light, which can also damage seeds

To avoid these problems, make sure the seeds are really dry and clean. Then, on a dry and sunny day, place the seeds in an air proof container.

To reduce moisture problems, add wood ash to the bottom of the container (of course, wood ash which has already cooled). Milk powder or other very dry seeds can be used as a substitute to absorb excess moisture.

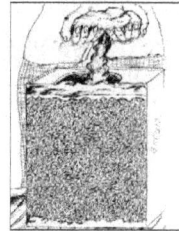

Reducing Insect Problems

The most common problems of insect damage to seeds can be avoided by using the following simple methods.

- Wood ash. Coat the seeds lightly in wood ash, and add more ash to the top and bottom of the seed storage container.

- Don't use ash from rubbish fires

- Neem. Add a 1 cm layer of neem leaves at the bottom and top of the seed storage container. Bay leaves or guava leaves may also be used

- Tobacco. Only use old and dry tobacco. Add a 1 cm layer of tobacco at the bottom and top of the seeds

- Gamal. Add a 1 cm layer of gamal leaves at the bottom and top of the seeds

- Cold temperatures. In places where it gets very cold at night, place the seed container outside every night for one week. Bring the container inside again every morning. This will kill insects such as weevils (a small white grub / worm)

- Salt. A small amount of salt mixed in with the seeds will also help to control pest problems

- Smoke. Smoke is a preservative and pest repellent. You can hang corn, seed pods and even whole plants above a fire to dry, and at the same time this will provide protection from pests

- Oil. Larger seeds can be coated with coconut oil to kill insect eggs. Pour a little coconut oil into a large container, add the seeds, cover the container and shake until all the seeds are coated in coconut oil. Small seeds cannot be treated in this way using ash and tobacco to protect seeds during storage

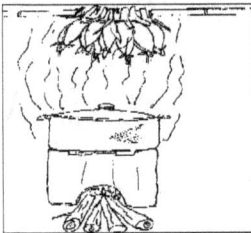

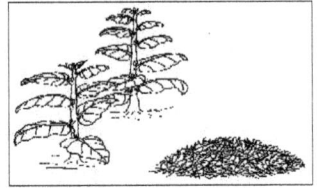

using ash and tobacco to protect seeds during storage

5.5 CONTAINERS FOR SEED STORAGE

Containers for storing seeds could be just about anything that can be used to safely store seeds. Keep the containers in a cool, dry and dark place. Protect from animal disturbances, and check the seeds fairly often to make sure there are no problems.

Seed saving containers can be made of woven bamboo, which has already been treated. You can coat the bamboo in tree resin, coconut oil or wax and then dry in the sun.

Tin cans and glass jars that have lids may also be used as seed saving containers. As well as plastic bottles and old film canisters, but be careful because mice can eat through plastic. Plastic bags can be used, but only if there are no other containers available, and they need to be placed in another container to protect them from animal damage. A large container with a good lid can be used to store many small bags of seeds. For larger seeds, biscuit

tins, old oil containers and large plastic containers will work well. Metal drums are also good seed containers, but can be expensive. Blacksmiths can make storage silos. Silos can be used to store corn, beans and rice seed in large amounts

protecting seeds from animals

If the storage containers are placed up high, wrap the container legs with a metal plate for protection against mice. Use all your imagination to trick mice, don't let the mice trick you!

5.6 LIVE PLANT STORAGE

Cassava, sweet potato, taro and yam, are all important food crops. The best way to store these seedlings for the next planting is to leave them growing in the ground, and only use the seedlings when needed.

Spices, like ginger and turmeric, are the same. If you need to store some roots, store them in animal proof containers that have air holes which are too small for mice to get through, this will allow air to pass through, which will help to prevent rotting. A light layer of wood ash will help to protect against insects and mice that will try to feed on the roots.

Community Seed Saving Group

Creating a community seed saving group is a great way to share excess seeds and increase seed variety for every group member. Within this group, members can also buy, sell, or trade seeds to introduce new plant varieties.

A community seed saving group is like a seed and planting material bank. The group collects and stores the best seeds and planting materials. These seeds are stored for the future, to grow, to trade, or to sell.

The whole process of saving seeds and distributing them will be much easier by working within a group.

5.6.1 Seed Exchange

Excess seeds can be exchanged with other people or groups. This will support increased plant variety for every person.

5.6.2 Seed and Plant Selection

Collect seeds from the healthiest, most disease resistant plants or from plants within the community. Generally, only 5-10% of community crops need to be left for seed collection. Members who grow plants for seed collection can be given compensation by selling or trading those seeds within the group.

Besides keeping group seed stock, it is important to find out how plants grow well. For example, suitable plant varieties, pest predators, amount of water and sunlight needed, and so on.

5.6.3 Seed Collection and Drying

The task of collecting and drying seeds is easier and quicker if it is done with the support of the whole group.

5.6.4 Seed Drying Room

A seed drying room is a room where most of the seeds are dried, especially during the wet season. This room must protect seeds from rain and can use smoke or heat to dry seeds. Give special attention to the room temperature and make sure the room has good enough ventilation.

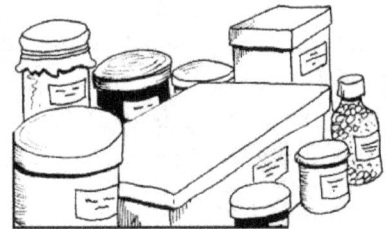

5.6.5 Seed Storage

A large community room or an agreed place can be used to store seed. Compiling containers for large amounts of seed is much easier and less expensive if done through a community group. Permanent seed containers or silos can be ordered or made, and can be used to fulfil the needs of the whole group.

5.6.6 Seed Supply

All seeds that are saved by the group should be used wisely. The seeds should be distributed evenly among the group members so that every member has enough seed for their own land. Every member that who receives seeds must give something in return to the group. This could be seed products, labour, manure, compost, land, storage containers, and so on. If there is excess seed, some can be kept in case of seed shortages during the next season. If possible, always save enough seed for one more crop season.

5.6.7 Seed Garden

A community seed garden can be made specifically for producing seeds. This garden will provide high quality seeds, because the seeds are taken from the best plants, and it will make it easier to reduce chances of cross pollination.

5.6.8 Seed and Planting Material List

A list will help group members know what materials the group has available. What is available could be seeds, plants, and planting materials. This list also provides information for people outside of the group who are interested in buying or trading. Other benefits of having this list include:

- ◉ It helps to identify the best place to grow each type of plant

- ◉ To identify the differences in types of plants

- ◉ It can be combined with other plant lists to form a district or national plant list

- ◉ It helps to assess what the community can produce and what needs to be introduced

- ◉ It helps to keep local plant varieties in the ownership of the community

If there are two or more types of the same plant, for example tomato, write these separately as two different plants with different names, for example round red tomato and bell shaped yellow tomato. This is because different types will have different amounts of produce, different disease and insect resistance, different time of fruiting, and even different eating quality.

The list includes:

- ◉ Plant name: Local name, Bahasa name, botanical or Latin name (if possible)

- ◉ Description: Plant description

- ◉ Plant size / shape

- ◉ Time of fruiting: How long after planting the seed will the plant produce fruit or leaves

- ◉ Consumption quality: Is the plant considered good to eat

- ◉ Susceptibility: What insects or diseases often harm the plant

- ◉ Uses: What are the plants uses, for example as medicine, building material, etc

Plant name	Plant name	Plant name	Plant name	Plant name	Plant name	Plant name
tomato	fruit	oval fruit	3 months	good	fruit pests	food, natural pesticide
yellow passion fruit	fruit	vine, oval yellow fruit	1 year	very good	pests attack seedlings	syrup, shade plant

5.6.9 Seed Testing

Seeds can be tested to find out how many will grow. When testing seeds for personal use, place the seeds in a container of water. The seeds that sink are the good ones to be planted, the seeds that float are the bad ones and should be thrown out. Usually, almost all of the seeds will sink.

For seeds that will be sold or exchanged, it is better to test them first to find out what percentage of the seeds will germinate and grow. This viability rate can then be written on the packets. There are a few methods to test this, one is to count a number of seeds (for example, 50 bean seeds), then plant those seeds and count how many grow (for example, only 40 seeds grew). Divide the amount of seeds that grew by the amount of seeds that were planted to find the percentage of beans that grow (40 : 50 = 0.8). This number is then multiplied by 100% (so the result is 80%). So, the viability rate of these seeds is 80%.

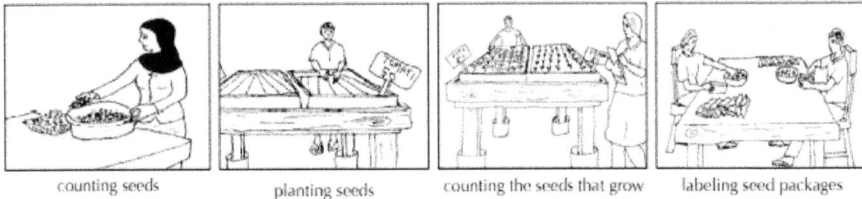

counting seeds   planting seeds   counting the seeds that grow   labeling seed packages

When conducting this test it is important to make sure the soil mixture used is of the best quality. Take good care of the seeds and protect them from pests, such as snails and ants. The test must continue until the seeds have passed the germination phase. The seedlings can then be planted in the ground.

5.6.10 Exchanging and Selling Seeds

The seeds can be packaged to exchange or sell within the community, or between other groups and towns. Selling or trading seeds requires a consistent supply of seed. Planting materials can also be exchanged, sold or bought. Make sure to test products before selling them, to ensure that all products are high quality.

5.6.11 Community Nursery

A community nursery should be made to support the common needs of the community group. Counting seeds planting seeds counting the seeds that grow labeling seed packages

5.7 MAKING A PLANT NURSERY

A plant nursery is essential in providing the best environment for plants when they are still young and fragile. Like children who need special attention when they are still young, so do plants. A healthy strong seedling will grow to become a healthy productive plant. The early stages of a plants life will determine how well it will grow in the future.

A nursery can be made any size you need, it can be small, the size of one garden plot with a coconut leaf roof, or large and managed by the whole community.

5.7.1 Some benefits from making a plant nursery include:

- Makes planting, watering and maintaining seedlings easier because everything you need is in one place

- Provides seedlings with protection from hot sun, hard rains, strong winds and animal disturbances

- Allows seedlings to grow healthier, because there is enough healthy soil and nutrients available

5.7.2 Plant Nursery Location

The nursery is the heart of the garden and needs attention every day. The nursery should be located close to the house and close to the garden. The nursery needs watering almost every day, so it is best to be located close to a water source.

Trees can be used as shade. However, be careful because too much shade can cause problems in the future, because the seedling will be too weak. Legumes, like sesbania and eucalypts, are good trees for a nursery, because they will still allow some sunlight to pass through. Trees like mango and avocado are not very good to use because they are too dense.

The best situation will allow morning sunlight in the nursery, and provide shade during the middle of the day and afternoon, because this is when the sun is hottest. Protection from strong winds is also needed, because strong winds will slow the seedlings growth. But, some gentle wind blowing through the nursery is very good for the seedlings.

5.7.3 Designing & Constructing the Nursery

Every nursery will be constructed differently to fit different needs and different construction materials. The following examples can be used or you can come up with your own nursery designs. Make the nursery design so that it lasts as long as possible. The nursery should have different areas which receive different amounts of sunlight. If possible, there should be 3 areas, which are for:

- ⊙ Small seedlings or fragile plants, which are still weak and need extra protection from hot sun and heavy rains.

- ⊙ Larger seedlings, which don't need a lot of protection, but do need enough sunlight.

- ⊙ Plants in the process of 'hardening' before planting. These plants need full sunlight in preparation to face the conditions where it will later grow. Larger plant seedlings need 3-4 weeks to 'harden' and small vegetable seedlings need 1 week to 'harden' before planting in the garden Nurseries on top of a para-para (bamboo frame that allow water to seep through) or waist high table will provide protection from animals, such as snails, ants and other insects. Also, nurseries will be easier to manage, because you won't have to bend over all the time, which is not fun and puts stress on your back. Remember, think smart, not hard!

Larger nurseries will be easier to construct and manage if a group of people are involved. This group could be a family, community group, school, or religious organization.

Every person involved will benefit more from the work they do and will save on production costs. Nurseries can be made separately in different places, or all the seedlings can be grown in the same place and divided up when it is time to plant in the garden. Or, use a combination of both.

Larger community nurseries can be made for reforestation needs.

5.7.4 Constructing the Nursery Building

For the main frame, use strong long lasting materials, like eucalyptus wood. Some types of long lasting bamboo can also be used for the frame, but some types are not very strong and will rot within 1-2 years. Bamboo that is harvested and treated correctly will last longer. (For more information about bamboo, see Module 8 – Forests, Tree Crops & Bamboo).

Roof materials can be bamboo panel, woven coconut / palm leaf, or grasses tied in thin clumps so that some sunlight can still pass through. Fence materials can be bamboo or wood, palm leaf, or any other available materials. Or you can even make a living fence.

In the mountains, seedling will grow better if they are raised off the ground, about waist high is best. This is because at night it becomes very cold. Cold temperatures can damage and even kill seedlings. Some other ways to deal with cold temperature are by planting some trees near the nursery or covering the ground with a layer of mulch made from coffee / rice husks, about 10-15 cm is good. However, a layer of husk mulch could promote fungus growth in areas with warmer temperature.

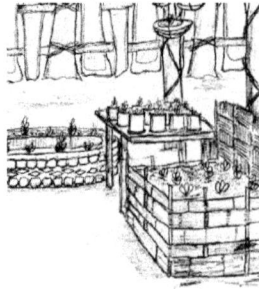

Seedling Boxes and Containers

**making containers
from banana trunks**

Seedling containers are easy to make and are good for growing many varieties of vegetable and tree seedlings. These containers need to be made deep enough for roots to grow long, not grow around in circles. Tree seedlings can be transferred to the containers when they are about 1 month old (about the time when four leaves have grown). Height and drainage is very important. Choose the size of container that fits your needs and the materials available. These containers are generally made of wood or bamboo. If using bamboo, the outside of the bamboo should be facing up to provide better water drainage. Many different containers can be used as seedling containers. All seedling containers should have drain holes at the bottom.

Seedling containers can be made from:

**making containers
from banana leaf**

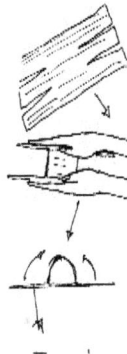

- Cans, baskets, used drink / food containers, and other used materials
- Coconut husks
- Bamboo
- Banana leaves, must be 1 finger width at the base to be able to hold water
- Banana trunk / bark
- Woven leaves
- Poly bags are the easiest containers for nurseries with lots of trees. They do cost money, but save a lot of time and energy

5.7.5 Soil Mixtures

Soil mixture for seedling boxes and containers is different than soil in the garden. It is important to make sure that the soil used will allow the plant roots to grow easily and water to drain easily (not stay stagnant), as well as supply enough ready to use nutrients for the seedlings. All plant seeds contain the food it needs for the first few weeks of growth. For best results, use different soil mixtures that the plant will need following this period.

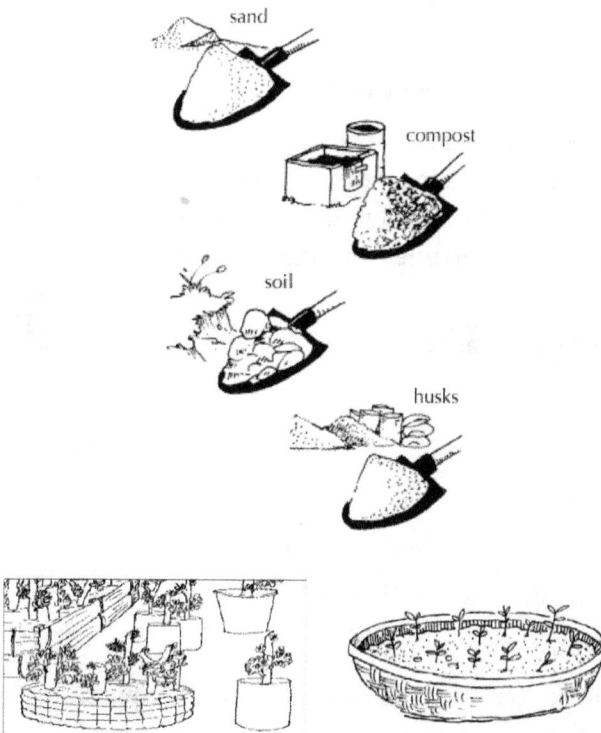

5.7.7 Soil for Cuttings and Seeds

Plant cuttings and seeds only need a small amount of nutrients during their first month of growth. In fact, too much nutrients will cause problems. Some examples of soil mixtures that are best for plant cuttings and seeds:

- ◉ 25% compost / dry manure
- ◉ 25% soil
- ◉ 25% sand
- ◉ 25% composted rice / coffee husks

Or:

- ◉ 50% sand or composted rice / coffee husks
- ◉ 25% compost / dry manure
- ◉ 25% soil

Or:

- ◉ 25% compost
- ◉ 50% sand
- ◉ 25% composted rice / coffee husks

A handful of wood ash can also be added toimprove the soil and balance the soils pH levels.

5.7.8 Soil for Long Term Trees and Plants

Plants that have been planted in containers need more nutrients to grow, especially long term plants. More compost or dry manure can be added to the soil mixture used. Some examples of soil mixtures for long term trees and plants:

- ◉ 30% compost / dry manure
- ◉ 30% soil
- ◉ 30% sand
- ◉ 10% ash or husks

Or:

- ◉ 50% compost / dry manure / husks
- ◉ 50% soil / sand

You can even make your own soil mixture, what is important is that we understand the functions of the following materials:

- Sand provides good drainage and aeration for easy root growth

- Coffee / rice husks also provide drainage and aeration, and can be composted before being used in the soil mixture

- Compost and dry manure provide nutrients. Don't use fresh manure, because this could burn the seedlings

- Liquid compost is good to use for plant seedlings over 1 month old

Fill the bottom of the seedling containers with a layer of small rocks, about a 3 cm layer is good, before adding the soil mixture. This will improve water drainage.

5.7.9 Fungus Problems

In the wet season, seeds and seedlings can become infected by fungus in the soil. This is a common problem which can cause seeds not to grow and young seedlings to rot. If this happens, there are two solutions:

- Reduce the amount of manure and compost in the soil mixture. For garden soil, fungus and bacteria are beneficial, but are not for nursery soil

- Before planting the seeds, pour boiling hot water over the soil in the container. Boiling water will kill any fungus in the soil mixture. Wait till the soil is cool again to plant the seeds

5.8 NURSERIES

5.8.1 Planting Seeds

Small seeds should be planted about one finger knuckle deep, while larger seeds should be planted about two finger knuckles deep. Plants which grow better in nurseries include cabbage, tomato, green vegetables, spinach, eggplant, capsicum, onions, chilli, cucumber, and okra. Plants which grow well if the seeds are planted directly in the garden include pumpkin, corn, beans, peanuts, radish, sunflower, squash, and melon.

However, almost all plant seeds will grow well if planted in a nursery.
It is good to label each seed planted. On the label, write the name and date planted. This is very useful, especially for large scale nurseries and community nurseries.]

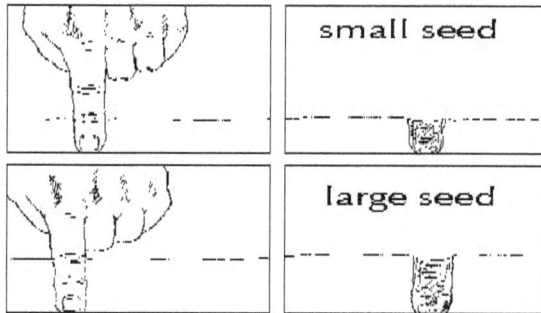

It is best to try and transfer seedlings from the nursery to the garden at the beginning of the wet season. Estimate the time needed for seeds to grow ready to be planted in the garden, for example:

- ⊙ Vegetable plants usually need 3-4 weeks from seed to planting in the garden
- ⊙ Fast growing trees need 2-3 months
- ⊙ Slow growing trees need 3-5 months

5.8.2 Planting Seeds Directly in the Garden

Some types of plants will grow much better if planted directly in the garden, for example carrots. However, these plants need special attention as they grow.

5.8.3 Steps for direct planting:

- ⊙ Dig the soil well before planting seeds. Add some sand if the soil contains too much clay
- ⊙ Water the ground
- ⊙ Plant the seeds close to the surface, then spread a thin layer (about ½ cm) of sand or soil
- ⊙ Water again, then cover the ground for 1 week to keep the ground moist
- ⊙ If there is no rain, water a little each day and then cover again
- ⊙ When seedlings start to grow, remove the cover and continue to water every day or every two days, for as long as two weeks

All vegetables and plants that are grown by root cuttings will grow better if planted directly in the garden, examples are: Sweet potato, potato, cassava, water cress, taro, garlic, ginger, and bamboo.

Collecting Young Seedlings

Sometimes the easiest method for growing trees, and even some is by collecting their young seedlings. These seedlings often ca parent trees. The seedlings should be collected when they are as 5- 10 cm in height is best. This will reduce stress and root dama of removing them.

collecting young seedlings

The process of collecting seedlings must be done very carefully, by digging them up, don't just pull them out. If the seedlings are larger than 20 cm, prune them back to 2 or 3 leaves high. Water the seedlings first, or collect them after rain to reduce root damage and plant stress. Replant the seedlings into containers, leave them in a shaded place for one week, then put them in the nursery and handle them the same as the other seedlings. If you want to plant them directly in the garden, give them shade for one week after planting. For vegetable and flower seedlings, just 3 days will be enough shade.

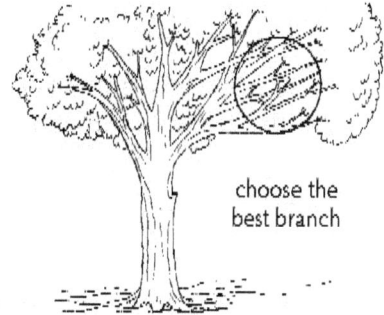

choose the best branch

5.8.4 Plant Propagation

There are many trees and plants that can be easily propagated. Some propagation techniques will be explained in the following, such as by using cuttings / branches, roots, aerial roots, and grafting techniques.

5.8.5 Branch Cuttings

Follow these steps to propagate plants using branch cuttings:

- Choose the best branch for propagating, usually aged about 1-2 years, with hard wood, brown color, but not tough and old.

- Cut the chosen branch with a sharp knife, so that both ends of the branch are clean. Make sure there are at least six growing buds on each branch. These growing buds are where new roots and leaves will grow from. Cut the top of the branch at an angle so that water will not sit on top, which could cause fungus and disease.

- Place the branches in a bucket of water until they are ready to be planted.

- Prepare the planting area. This can be the plant nursery, containers, or directly in the garden. If planting in the nursery, use the same soil mixture used for planting seeds.

- When planting, make sure that the growing buds face upwards. Also make sure that there are 3 growing buds below soil and 3 above soil.

- Water every day until new leaves grow. Then, water only twice a week. If planting directly in the garden, water every day if there is no rain and provide shade until new leaves grow.

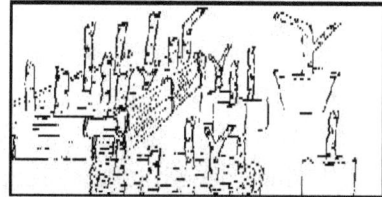

5.8.6 Root Cuttings

Follow these steps to propagate plants using root cuttings:

- ◉ Water the plant before cutting its roots

- ◉ Dig the soil, first from the side under the plant, then straight down through the plant to cut and separate the section of root that will be removed.

- ◉ Remove the section, be very careful not to damage the roots.

- ◉ Carefully replant the plant root into the ground or in a container, and water well.

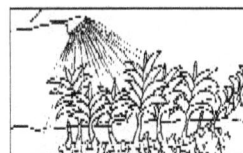

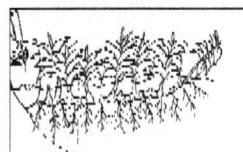

5.8.7 Marcotting / Aerial Roots

Propagating plants by creating areal roots is a commonly used practice. This method is good to use and quick; it creates new plants that if planted from seed / seedlings could take 2 or 3 years to grow as large as the aerial root plant. The steps are:

- ◉ Choose a strong, productive and disease and pest resistant plant

- ◉ to make the marcott. This is very important, because the new plant that will grow will have the same characteristics as its parent plant.

- ◉ Choose a healthy branch, positioned in the shade. Peel off the outside layer of bark, about 10 cm (middle finger length) of the branch.

- ◉ Cover the part of the peeled branch with healthy soil and wrap it in plastic. Tie both ends of the wrapping, and if needed the center as well. The soil inside the wrapping should stay cool, so if needed use two layers of plastic.

- ◉ Make sure the soil stays moist, and check it regularly. Leave for 3 months.

- ◉ After 3 months, there will be a lot of roots growing. At this time, the branch is ready to be cut. Cut it below the marcott, it is best to use a saw to avoid damaging these roots.

- ◉ Carefully, remove the wrapping. Put the roots in water until it is time for planting. Remove about 1/3 of the branches and leaves. Plant into a pot and place in a

shady place until new growth starts. When this new plant is established, move it into a place that receives enough sunlight. When the tree becomes strong and hard it is ready to be planted in the garden.

5.8.8 Grafting

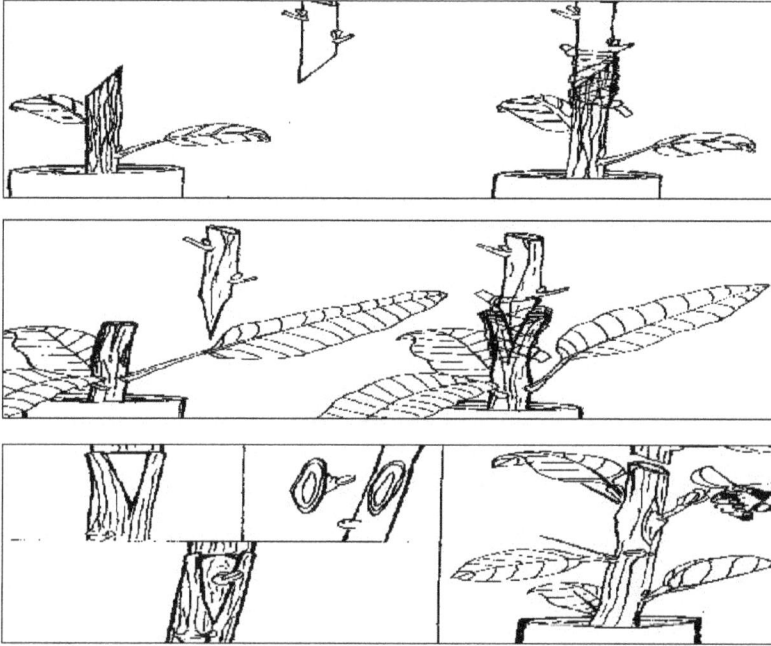

Grafting is a technique used for fruit and nut trees to improve quality, productivity, and save time between planting and harvesting.

This technique is difficult and requires a lot of practice. A simple explanation is that a branch from one healthy and quality fruit tree is attached to the stem of another tree of the same type. For example, mango to mango, orange to orange, coffee to coffee.

A full explanation would be too long to include in this book. If you are interested in learning more, information is available from the government agriculture department or NGOs working in the agriculture field. Remember that grafting is not the only way to produce high quality, productive fruit trees. Soil improvement, water storage, organic fertilizers, mulching and maintenance are all essential factors to increase production. Grafting is one important technique to help improve future quality and production, but only if the essential factors have already been addressed.

5.9 NURSERY MAINTENANCE

5.9.1 Watering

- ◉ Plants in a nursery need watering almost every day. Be careful when watering young seedlings, too much water at one time can kill young seedling because they are still very fragile.

◉ For established plants in containers, watering frequency can be reduced. However, check them regularly to make sure their soil does not become dry. If the soil is dry as deep as one knuckles length, the plant needs watering.

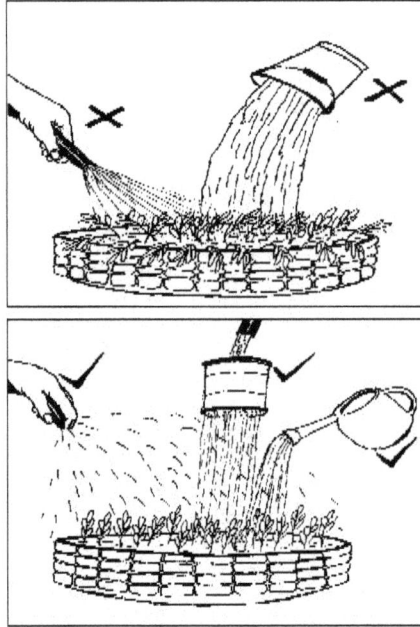

5.9.2 Fertilizing

◉ Seeds will grow stronger and faster if they are receiving enough nutrients. This will also make the plants stronger and healthier later on. Liquid fertilizer contains a variety of nutrients, and is good to use for seedlings. However, don't use liquid compost for seedlings under 1 month old. Dilute liquid compost with water before use. The amount of water used to dilute should be more than what is normally used to dilute liquid compost for use in the garden.

◉ This fertilizer can be used once or twice a week. Plants in containers will suffer if too much fertilizer is used in the soil mixture. It is better to give the plants only a small amount of fertilizer in the nursery, and more when they are already planted in the garden.

5.9.3 Transplanting Seedlings

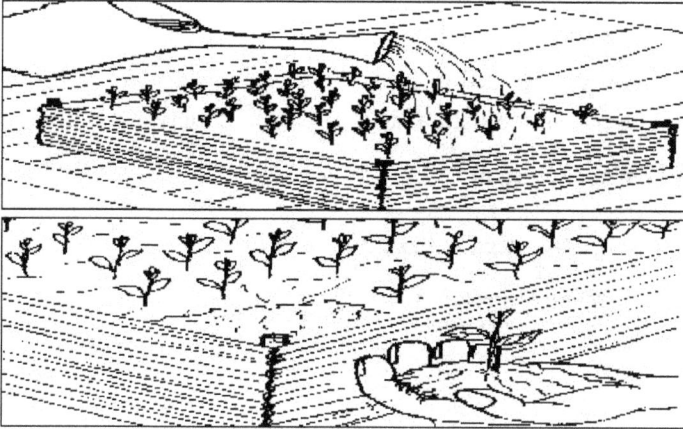

In the nursery, small plant seedlings will sometimes need to be transplanted into a larger container.

5.9.4 The safest way to transplant them:

- ◉ Water the seedlings well
- ◉ Dig the seedlings up using a small shovel or your hands. Don't pull the seedlings out by their stems!
- ◉ If there are many seedlings together, separate their roots very carefully
- ◉ Immediately replant into another container, with the roots pointing down.
- ◉ Carefully, water again

5.9.5 Weed and Pest Control

5.9.5.1 Weed Control

Weed control is very important in nurseries. The weeds will compete with the seedlings for food, and hence slow down their growth rate. Continually remove any weeds that grow around the seedlings.

In a garden nursery, apply a thin layer of mulch to stop weeds from growing. This layer of mulch can be combined with a layer of plastic in between plants.

5.9.5.2 Pest and Disease Control

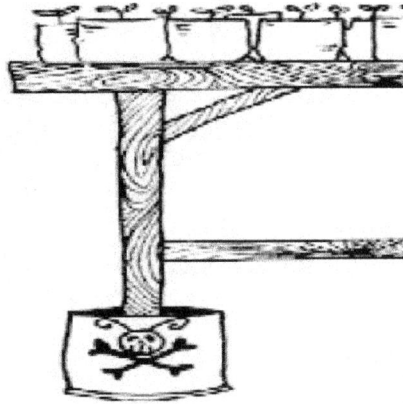

Pests and disease can spread easily and quickly. The best solution is to prevent pest and disease problems before they occur by:

- Reducing plant stress as much as possible by protecting them from hot sun and allowing gentle winds to pass through the nursery
- Using a suitable soil mixture
- Watering enough
- Fertilizing enough, but not too much
- Raising seedling containers and boxes off the ground
- Preventing insects from reaching the seedling boxes, for example by placing table legs into a can of water to stop ants, snails and slugs from climbing up.

You can reduce the chances of pests or diseases spreading from plant to plant by combining a variety of plants together, or keeping plants in smaller groups, rather than all together in large groups. If plants are attacked by pests or disease, you can:

- Spray them with a natural pesticide. (For more information about natural pesticides, see Module 9 – Integrated Pest Management)
- Change the conditions of the environment, for example if plants are suffering from fungus or mildew, try providing more sunlight and wind
- If other solutions don't fix the problem, remove and burn diseased plants

Transplant seedlings into the garden before their roots grow too large for the containers. If plant roots grow too large they will grow around in circles and get stuck there. This is called 'bound roots'. It will slow down the plant's growth rate, and can even cause plants to die. Plant roots are a very important part of a young plant. Healthy and strong roots will produce a healthy and productive plant or tree. There will always be more roots than leaves when the plant is still young. If plant roots do get stuck, you will have to trim off the outer roots to stimulate new root growth. If you trim off some roots, you should also trim off some branches. Make sure plant roots are not growing out of containers and into the ground. If roots are only just starting to emerge from containers it is usually not a problem, but if too many roots grow out of the container and into the ground they will need to be trimmed off, which can cause the plant damage or even kill it. One way to avoid this happening is by placing seedling containers far above the ground.

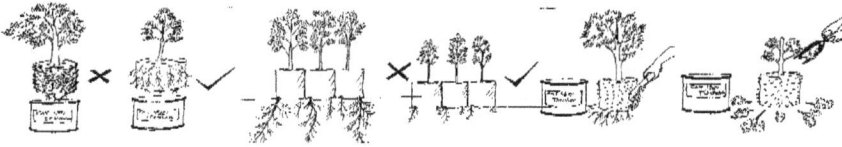

Hardening Plants

All plants that are grown in nurseries should be 'hardened' before they are transplanted in the ground. This means preparing the plant for the conditions where it will later be planted. Hardening a plant could mean leaving the plant in the sun for a time, except for plants that need to be planted in shade, like coffee and vanilla. Hardening plants is very important because it reduces plant stress during planting. If the plant is not hardened first, it may stop growing or be stagnant for a few weeks, and could even die because of experiencing too much stress. Another technique to reduce plant stress is to provide the plants with shade for about a week after they have been replanted in the garden. The more plant stress is reduced, the better it will grow. This is the same as people.

6

HOME & COMMUNITY GARDENS

6.1 GOOD NUTRITION

- ◉ Planting a range of vegetables, fruits and grains is important for providing family nutritional needs, especially for children.

- ◉ Good nutrition is especially important for pregnant and breast feeding women. Other family members need to help make sure that pregnant and breast feeding women are getting enough of the best foods possible.

6.1.1 Some benefits of good nutrition include:

- ◉ Reduced health problems
- ◉ Faster recovery after sickness
- ◉ Children grow better
- ◉ A longer lifetime
- ◉ More energy for activities
- ◉ Increased ability to learn and concentrate. This is very important for children who are still in school. Better food will create smarter people We need to eat a variety of foods to be healthy.

Every day we should eat vegetables, fruits, eggs, and meat, as well as beans and grains. A wide range of healthy vegetables grown at home will provide many vitamins, minerals, proteins, energy and oils.

6.1.2 Sources of Nutrition from the Home Garden

- Vitamin A: Good for eyes, examples are taro leaves, sweet potato leaves, cassava leaves, pumpkin leaves, cabbage, green vegetables, carrot, mango, banana, papaya.

- Vitamin C: Good for body health, examples are papaya, citrus, tomato, pineapple, guava, tamarind.

- Protein: Strong bones and muscles, examples are peanut, beans, peas, yam, watermelon seeds, banana tuber, moringa seeds, candle nut.

- Carbohydrates: For energy, examples are rice, corn, sweet potato, cassava, taro, potato, avocado, coconut (old), jack fruit, bread fruit, sugarcane.

- Fats and oils: Good for skin and hair, examples are avocado, milk, chocolate, peanut, candle nut, cashew nut, soybean.

- Iron: Good for growth, strength and stamina, examples are mustard, spinach, green vegetables, banana tuber, cassava, sweet potato leaves, dried beans.

nutritional foods
are needed
every day

Other vegetables such as eggplant, squash, pumpkin, cucumber, onions and radish, and fruits such as watermelons, bananas, apples and much more, also provide a lot of vitamins and minerals. Some types of trees provide very nutritional leaves, roots, sap, trunk and bark.

Meat, fish and egg provide a lot of protein, iron and oils. If possible, eat these every day. Dried beans, tempeh and tofu are also high in protein.

Mushrooms are very nutritional, good for health, and provide protein as well as many vitamins and minerals. Mushrooms can be collected in the wild, or by using manure, liquid compost and mulch they can be grown in the garden and vegetable plots. This is because mushroom spores (or seeds) live in and are spread by manure, compost and mulch.

Spices and herbs, such as chili, ginger, garlic, pepper, coriander and basil are also important to eat for the bodies health and are good to use to fight some sicknesses.

Traditional medicinal plants, like aloe vera, kumis kucing, samiroto and daun sembung, can also be planted near the house, in between flowers and vegetables.

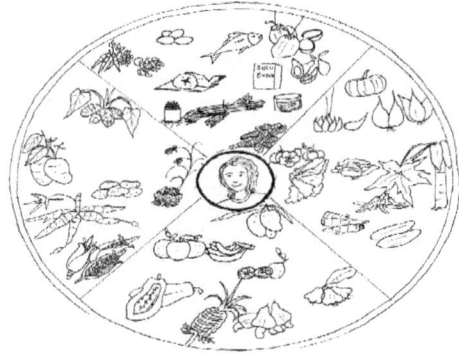

All plants the family needs can be planted ourselves, which means we are fulfilling the families needs at a low cost. Excess produce can be sold or exchanged.

When cooking, remember that many vitamins are lost if vegetables are cooked for too long or if water used to boil vegetables is thrown out.

6.1.3 Nutritional foods are needed every day Designing a Garden

There is a lot of knowledge about agriculture in Indonesia now, which is still growing and developing. Improving food production depends on the knowledge willing to be shared between communities.

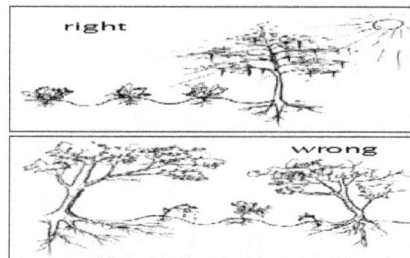

This module uses a lot of this knowledge and adds to it new techniques, which use local materials to fit local needs.

6.2 GARDEN LOCATION

6.2.1 Sunlight

Plants need sunlight to grow. Plants use sunlight and change it into food through a process called photosynthesis.

Almost all plants prefer to receive full sunlight. However, some plants like spinach, beans, cabbage, cucumbers, lettuce, potato, pumpkin and other green leaf vegetables, can still photosynthesize well under some shade.

Don't plant tall growing and thick leaved trees, like mango and jack fruit, near vegetable plots. As these trees grow larger, they will block out the sunlight.

Some other types of trees can be planted near vegetable plots, such as banana, papaya, and legume trees like acacia and casaurina. Don't plant too many trees or shade plants, use them only as needed.

6.2.2 Water

Water is always needed for planting any type of vegetable, not only during the dry season, but also during the wet season in some areas that are particularly barren. So, gardens should be close to a water source or have good irrigation. Irrigation can be made using bamboo, metal or plastic piping. Storing irrigation water in a tank or drum closer to the garden will help to provide a continuous water supply. All stored water should be covered to prevent mosquitoes from breeding. Covering water will also help to reduce water loss due to evaporation.

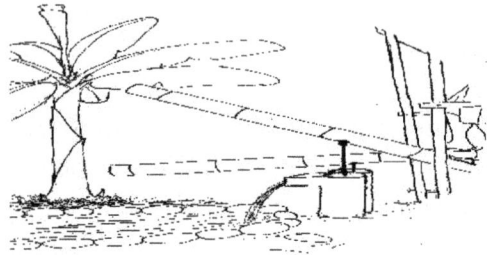

Use gravity to help make irrigation, this is easier and less expensive. By using gravity, water can be run from higher places to lower places. Hand pumps are also good for bringing water up from underground sources.

Any irrigation must be designed in cooperation with other water users. If a community group is formed, tanks, pipes and hand pumps will be cheaper to buy and much easier to maintain.

6.2.3 Soil

The garden location should have healthy soil, and also be close to the house and a water supply.

Almost all soils can be improved quickly with good techniques and by regularly using mulch and compost. Soils that contain a lot of clay or are waterlogged need time and specific techniques to make them productive. Maybe it will be more productive to use these areas for something else, such as for fishponds and water plants.

6.2.4 Wind

Vegetables, especially seedlings, need to be protected from strong winds, which can dry out the soil and reduce moisture in plants. Living fences and windbreaks will help manage problems associated with strong winds.

a windbreak system

6.2.5 Other Factors

6.2.5.1 Root Competition

Large trees have roots that spread out up to 2/3 the height of the tree and the same width as the tree. These roots will compete with vegetable plants for water and nutrients in the soil. Some trees, such as eucalypt, are especially competitive, so these trees should be either removed from the land or regularly cut back to reduce their roots size. The eucalypt tree also releases an oil (alelopati) from its roots which most other plants don't like. Thin leaf legumes, like acacia and sesbania, or smaller fruit trees, like banana, papaya and guava, are examples of suitable trees to be planted in the garden.

6.2.5.2 Distance from house

Having the house and vegetable garden close together will save time, energy and costs. Because of this, we must first decide which types of plants are going to be planted. Larger plants that don't need intensive management and are not for every day use can be planted further away from the house. While, plants which do need intensive care and can be used every day should be planted close to the house,

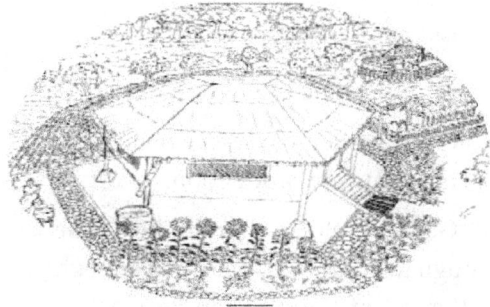

such as vegetables and bamboo. Gardens made close to the house will also receive benefits from house wastes.

6.3 PREPARING THE GARDEN

6.3.1 Garden Plot Design

In conventional agriculture, garden plots are generally made in long rectangular shapes and straight lines. These shapes actually are only suitable in low lands, while in higher areas where the land is sometimes more sloped, these can be very difficult to make. Isn't it true that we can find no square or rectangular shapes in nature?

Only commercially focused people are benefited from using this type of system, because they can count how many trees and plants they have. Try to think and act creatively, remember that beauty and natural patterns are also important. The easiest way to make garden plots is by following natural shapes, follow the natural shape of the land you are working on. Besides the shapes looking more interesting, pest problems will reduce and land use will be maximized. Working against nature increases the possibility of problems. Raised garden plots should always be surrounded by rocks, bamboo, wood or any other material that will:

- Hold the soil
- Hold more water in the soil
- Hold mulch
- Allow the soil to build up

Good garden plot design will improve soil quality. Improving the quality of the soil will also improve production.

Garden plots should be wide enough to hold water, but small enough so that all of the plot can be reached without being trampled. A width of ½ meter to 1 meter is good, or maybe 1½ meters if you have long arms. If the garden plots are often stepped on this

will cause soil compaction, which is not good. Garden plots should be designed with main pathways which can be used for bringing in compost and mulch, and for bringing out garden produce, as well as smaller pathways for access and to make garden maintenance easier.

During the wet season, the edge of garden pathways can also function as swales to collect and hold water. There are many other garden plot designs which function very well in dry lands and conserve water maximally.

6.3.2 Swales

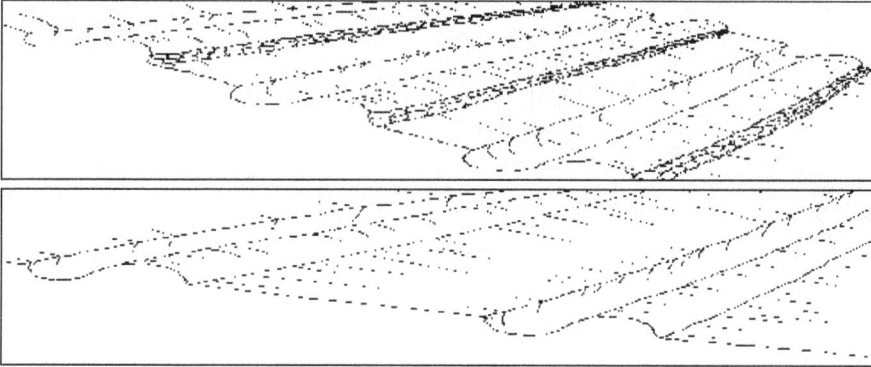

For areas with sloped land, swales are a great way to make vegetable gardens. This can even be used for small home gardens.

On steep slopes, swales will help prevent erosion, while still holding water and nutrients in the soil. Swales and terraces should be made following the shape of the land, so that if heavy rains come this will not create problems. For vegetable gardens, smaller swales are usually better. On steep slopes, make smaller swales about 1 meter apart. On gentle slopes, make the swales larger, about 2 meters apart. (For more information about how to make swales, see Module 8 – Forests, Tree Crops and Bamboo).

6.3.3 Terraces

Terraces are similar to swales because they are made following the land contour. Terraces cut into the land, and are usually stone or clay walls designed to hold the land in place. Terraces take more time, energy and cost more to make, but they will make the land very productive. Terraces are used in many countries and there is a lot of information about how to build and use them.

- *Try to make the edge of the swale or terrace higher using rocks or other materials. This will help hold more mulch, compost and water in the soil*

- *On steep slopes, make sure that heavy rains won't cause erosion or land slides. Use legumes to hold the soil in place, as well as for serving many other functions*

6.3.4 Fences

Fences are very important if you don't want pigs, goats and other animals eating all of your vegetables!

Remember that fences are multifunctional. Using a fence to separate two areas will save time, labour and resources. Planting a living fence will provide many more functions than a normal fence. Some of these functions include acting as a windbreak, trellising for vines, and for providing shade, animal habitats, and erosion control.

Living fences can be made from many different types of plants and trees, and can produce a range of products. Some products from living fences could include human food, animal fodder, mulch and compost material, medicines, wood, weaving material, nitrogen fixing legumes and natural insecticides.

Living fence materials: Leuceana, cactus, sesbania, moringa, tall grasses.

Other fence materials: Rocks, wood, bamboo, old fishing net, old tin roofing.

6.4 SMALL GARDEN NURSERIES

A garden nursery is important because plants need more care when they are still young. If seedlings are cared for carefully, the quality and size of vegetables will improve. A small garden nursery can be made from inexpensive and natural materials. You can also make a small movable nursery.

A nursery needs to have shade, healthy soil, and protection from animals, pests and disease. Don't ruin the land around the nursery by digging up soil for use in the nursery.

Following are some soil mixtures which are good for use in nurseries:

- 30% compost / dried manure, 30% soil, 30% sand, 10% ash / rice husks

- 50% compost / rice husks, 50% soil / sand

The soil needs to be combined with other materials. Sand and rice husks provide drainage, which makes root growth easier. Compost and dried manure provide nutrients and hold more water in the soil.

If seedlings are being planted directly into the garden, add rice husks, sand, compost and dried manure to help the seedling grow better. Also, build a temporary shade structure for the first 3-4 weeks after planting.

You can also just use containers as a nursery substitute. Many old materials can be reused as seedling containers.

6.4.1 Garden Additions

The garden can also be planted with small fruit trees, perennial plants, legumes and flowers. This will protect the garden from strong winds, provide food for humans and animals, and materials for making mulch and compost. Pollinators and pest predators, like birds, bees, spiders and other insects, will also be attracted into your garden. Increasing pollination of fruits and vegetable flowers will produce more fruits and vegetables per plant. Pest predators will feed on insects and pests, which will reduce the number of pests in your garden.

Flowers and herbal plants will add beauty and fragrant smells to the garden, as well as providing many other benefits.

6.4.2 Ponds

Ponds will provide many benefits in different ways. Ponds can produce fish, vegetables and materials for making mulch and compost.

Make one or two ponds near the garden area, the pond will attract frogs, small lizards, insects and birds, which will all function as pest predators in your garden.

- ◉ Excess water in the wet season can be stored in ponds to prevent water laying stagnant on the ground
- ◉ To manage the problem of mosquitoes laying eggs in ponds, add a handful of neem leaves to the pond once every three months. Neem leaves will help stop mosquitoes from breeding, but won't harm other pond creatures. Frogs, lizards and fish, especially tilapia fish, will feed on mosquito eggs and larvae

6.4.2 Garden Maintenance

6.4.2. 1 Providing Plant Food

Garden plots should be covered with compost at least 2 weeks before planting. Compost can be lightly turned into the soil if nothing has been planted, or just leave the compost laying on top of the soil. Add more compost 1 or 2 weeks after planting. Make sure adding compost does not disturb the plant roots. After composting, add a thick layer of mulch on top. Liquid compost can be used on garden plots every 1 or 2 weeks, but make sure to dilute with water before use. There are many different ways to fertilize the garden. It is up to you to decide which method works best for your situation.

Use EM (Effective Micro-organisms) with other soil improvement techniques to increase results.

All of these techniques will improve soil quality, structure and nutrient content so that there is enough food available for plants to use.

6.4.2.2 Watering

- Always water early in the morning or in the late afternoon. Morning and afternoon is better for watering because watering at night could promote fungus growth, while if you water at mid day, water will evaporate before it can soak into the soil, so the water is just being wasted

- Making garden edges will help to hold more water in the soil. Use rocks, bamboo, wood or other materials to hold the soil in place

- Mulch will protect the soil from hot sunlight and prevent water evaporation. This will also reduce the soil temperature and the amount of water needed for each garden plot

- 4. Making windbreaks around garden plots will save a lot of water. Wind dries out plant leaves and makes them lose water, so the plant then uses more water from the soil. Less wind means plants need less water

- 5. Watering with pipes. There are many used water bottles around and burning these bottles causes pollution. One way of reusing these bottles is by turning them into watering pipes, so that they water deep into the soil. Bamboo can also be used as a pipe, especially for fruit trees

Some benefits of watering deep include:

- Water evaporation is reduced because water is released in the soil, not on top of the soil

- Water can be concentrated at the roots of each plant

- Only a small amount of water is used

- Watering pipes can also be used to give liquid compost to plants

- Garden plots which are dug low need less water than raised plots, especially for in very dry areas

6.4.5 Weed and Pest Control

6.4.5.1 Weed Control

Weeds are an easily available mulch and compost material, which can also be used as animal fodder. Weeds should be more understood as a benefit rather than a problem. Reuse weeds to help keep the soil healthy. However, removing weeds can take a lot of time, and some types of weeds do create problems if they are not controlled.

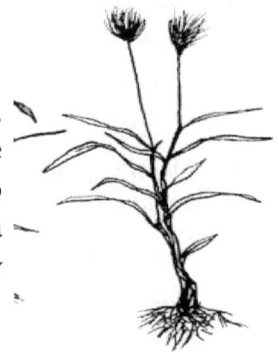

6.4.5.2 Some natural methods to control weeds:

- Continuously mulching the garden. Mulch stops sunlight from reaching the ground surface. When weed seeds grow they need sunlight to photosynthesize and keep growing, when sunlight is blocked by the mulch, almost all weeds will die. Try not to use weeds that contain a lot of seeds in the mulch because these may grow, and this will spread more weeds. If using quick growing grasses in mulch, make sure the grasses have been dried first so that they will not grow in the garden

- Use integrated planting systems. Vine plants and ground covering plants, such as pumpkin, beans, sweet potato and potato, can be planted under cassava, corn and other larger crops, to prevent weeds from growing. This same technique can be used for fruit trees or other tree crops

- ⊙ Use weed barriers. Make a weed barrier along the outside of garden plots to stop fast growing weeds. This weed barrier can be:

- ⊙ A space around the edge of garden plots which is always kept free of weeds

- ⊙ A small, thick living fence to prevent fast running grasses from entering the garden.

Lemon grass and other smaller grass plants can be used as a living fence weed barrier

- ⊙ Every time soil is turned, weed seeds are encouraged and are more likely to grow. Therefore, if you turn the soil less, fewer weeds will grow

- ⊙ Use animals as 'tractors'. This is a good way to remove weeds and their seeds, while fertilizing the land at the same time

- ⊙ Remove weeds before they produce seed. If weeds are removed when they are still young, the roots of vegetables will not be damaged because of weed removal

Lets create new weeds! These new weeds can be useful pants, which are intentionally planted to grow fast and spread easily. Choose a few types of vegetable, animal fodder or legumes that can function as weeds, it is important that these plants grow easily and quickly.

6.4.6 Pest Control

Pest control in the garden does not just mean exterminating pests. Controlling pests in a sustainable way involves using a number of techniques, from which the results will not be achievable from just using pesticides.

These techniques improve soil quality, encourage pest predators and prevent pests. If pesticides are still needed, use natural pesticides, not chemical pesticides. (For more information about pest management and recipes for making natural pesticides, see Module 9 – Integrated Pest Management).

6.5 PLANTING METHODS

6.5.1 Seedlings

Plants will react to damage or mistreatment, especially when still seedlings. Any damage caused will slow plant growth and reduce the amount of harvest.

6.5.2 For good seedling care, give special care to:

- Plant small seeds about 2 cm deep and larger seeds 3-4 cm deep. Don't forget to water seedlings every day

- If planting large seeds in containers, soak the seeds first to encourage faster growth

- Don't plant too many seeds in one pot. When the seedlings grow, they will need space for root growth. If planted too close together, plants will compete. Also, many roots will break as they are separated, and this will slow plant growth

- Increase the amount of sunlight seedlings receive for as long as one week before being transplanted in the garden. This technique is called 'hardening seedlings' and is used to prepare seedlings for stronger sunlight conditions in which they will later grow

- When planting seedlings in the garden, make sure they receive enough water

- Give special attention to plant roots. Make sure that plant roots are always facing down. Don't leave plant roots exposed to sunlight and avoid root damage

- Don't plant seedlings at mid day, when the sunlight is at its hottest

- Provide shade for about a week after the seedlings have been planted in the garden. A temporary shade structure can be made of anything, from legumes to woven coconut palm, or any other available material

6.6 SUCCESSION PLANTING

Don't plant all of your garden plots at once. By planting 3 crops of the same vegetable at different times, you will get 3 harvests. Even though harvests will be smaller, they will be extended and provide continuously. Also, less harvest will have to be thrown out or will go rotten. You can also plant different types of vegetable, which can be harvested at different times. Every type of plant needs a different amount of time to be ready for harvesting, so crop harvests will happen at different times.

6.6.1 Food Calendar

JANUARY		FEBRUARY		MARCH	
planted	harvested	planted	harvested	planted	harvested

A good technique for planning continuous food production is to make a food calendar.

◉ Make a list of all the vegetables and grains you want to grow. You can add illustrations to the list if you like

◉ Write down the planting times and harvest times

◉ Write out each month on the calendar, and list what was planted and what will be harvested each month

◉ If there are months that do not have harvests, consider:

- What else could be planted to be harvested in that month?

- Are there different types of plants which can be planted?

- Are there other techniques to increase harvest and make harvest times longer?

- What types of crops can be harvested continuously throughout the year?

If you have enough water supply, planting can be extended through most of the year. Mulch, compost and good garden design will help keep water in the soil for longer. This will extend the production period for crops.

6.6.2 Using Different Plant Growth Periods

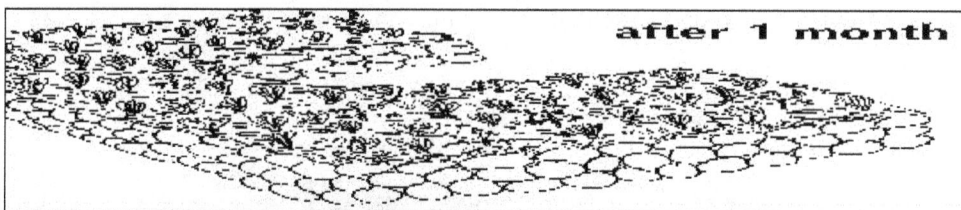

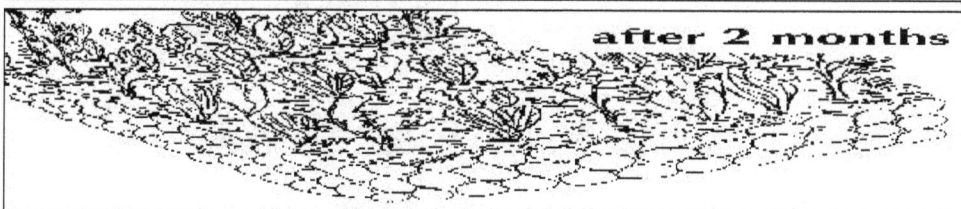

after 1 month

after 2 months

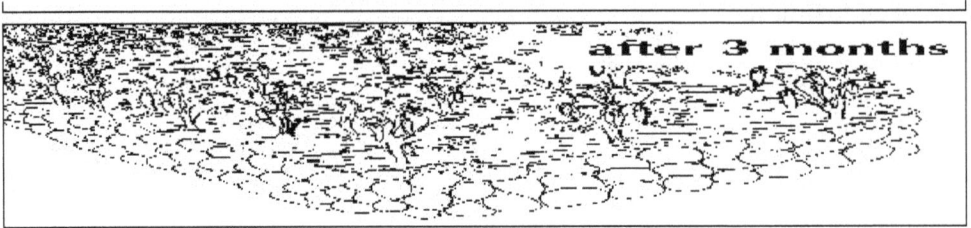

Every plant has a different growth period and will produce harvest at different times. You can use this knowledge to increase production in each garden plot. Lettuce, mustard greens and other green leaf vegetables grow quickly and can be harvested in 1 to 2 months. Eggplant, chilies, cabbages, capsicums, tomatoes and beans need 3 months or more to be ready for harvesting.

It is better if these plants are all planted at the same time, so lettuce and green leaf vegetables will be ready for harvesting before the other vegetables have grown large. Afterwards, there is still a following harvest of the other vegetables which take longer to produce. This means we will have more harvest times, with more crop variety.

Be careful not to disturb the roots of long term crops when harvesting the short term crop.

6.6.3 Using Different Plant Heights

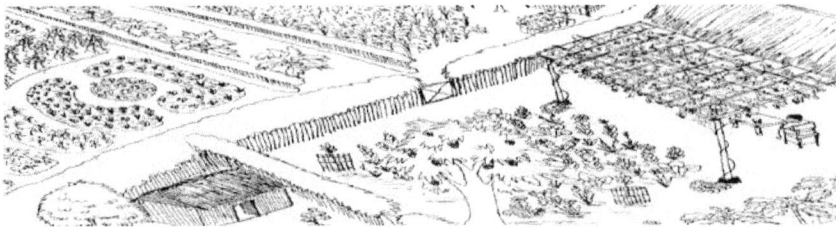

Plants that grow to different heights can be planted together to increase production amounts, while at the same time saving space in each garden plot. Make sure that smaller plants are receiving enough sunlight. Taller plants can be used as a place to grow smaller climbing vine plants.

6.6.4 Using Different Garden Plot Heights

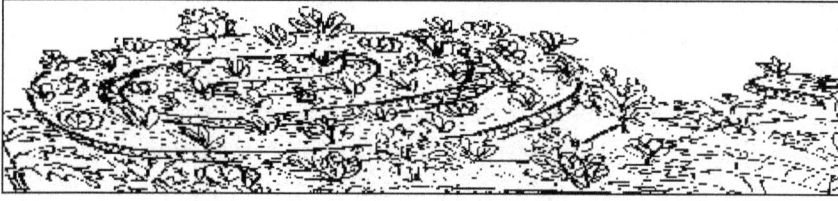

Using different garden plot heights can increase production and planting area. Different heights will allow more root growth and better access to sunlight.

Using swales on sloped land will provide more planting area and also provide different micro-climates. The bottom of swales are wet and sometimes full of water in wet season, so water plants, like kangkung and taro, can be planted there. The top area is drier and other crops can be grown there.

6.6.5 Crop Rotation

Different types of plants use different amounts of nutrients. Crop rotation helps to balance the amount of nutrients in the soil. Plant rotation will also help reduce pest and disease problems. It is better not to plant the same type of vegetable in the same garden plot twice in a row. All beans can be considered as one type of vegetable. Tomato, eggplant, potatoes and capsicum can all be considered one type of vegetable because they are all from the same family (solanaceae).

Once every two years, give each garden plot a few months of rest time to recover its stock of nutrients. During this resting time, add compost and mulch to the soil.

6.6.6 Crop Integration

Growing different crops together will reduce pest and disease problems, because pests will need more time to move from one plant to the next. Also, there will be less of each type of vegetable to be attacked by pests. Therefore, pest problems will be much easier to control. Some plants will benefit from other plants growing near them. For example, garlic helps to repel aphids (a very small pest, that in large numbers can damage tomatoes, capsicum, cabbage, green vegetables and other crops). Aphids don't like garlic. Therefore, by planting garlic near plants which aphids like, it will help to discourage and reduce the amount of aphids. Planting flowers and herbs in garden plots will attract insects, which will help with pollination, as well as increase the number of pest predators. So, this is also reducing pest problems.

Beauty is an important part of every garden. Integrating different types of plants together will make the garden much more beautiful and appealing.

- ◉ You don't have to plant vegetables and other crops in straight lines. Different patterns might even increase produce

⊙ Place long term crops, which don't need a lot of maintenance and will only be harvested once, at the back of the garden plot or in places which are difficult to reach. Place short term crops, which need more maintenance, and will be harvested over and over again, in areas of the garden plot which are easily reached. This will make gardening easier, and reduce soil compaction in garden plots

Following are some examples of vegetable combination which are commonly used:

⊙ Corn, pumpkins and beans

⊙ Tomato, garlic and basil. This combination grows well in smaller gardens and will help to protect each other from pests

⊙ Chilies and tomato

⊙ Sunflowers planted around the garden will help reduce pest problems

⊙ Cabbage, tomato and garlic

⊙ Carrot, onion, cabbage and lettuce

⊙ Cucumber, beans and peas

⊙ Sweet potato and taro. This combination works well for soils containing many rocks

Make a rock pile about 2 x 2 meters. Use large rocks, at least hand sized, so that there are many gaps in the pile. Around the rock pile dig a shallow pit, about 1 hand length deep. Add sweet potato and taro cuttings when filling in the gaps in the rock pile. Continue to add soil, rocks, and sweet potato and taro cuttings until the pile is 1 meter high or more. The result will be a pit or cave that can be used to grow sweet potato and taro coming outside of the pile. The rocks will protect the plants from mouse pests. Don't forget to add compost or fertilizers. Use cut banana stalk as mulch. This will help to keep the pit moist.

6.6.6.1 Integration with Animals

Plants need animal manure as their food, and animals need plants as well. This common need can go much further by integrating crops and animals together. This integration could be:

- Land use being rotated between crops and animals. Animals will clean weeds, loosen the soil and provide fertilizer after crop harvests

- For smaller gardens, chickens and pigs can be kept in movable pens to clean and fertilize the soil

- Vegetables can be grown at the bottom of a fish pond, which is dry during the dry season (if the pond is made of clay, and not cement)

- Vegetables can be planted along the edge of fish ponds (For more information about integrating animals and crops, see Module 11 – Aquaculture and Module 10 – Animal Systems).

6.6.6.2 Integration with Rice

Rice paddy borders can function as garden beds. Vine plants, such as beans, luffa, cucumber and pumpkin can be grown along these borders.

Rice crop and water plants can be grown together in wet areas. Water flowing through the rice paddies can be stored in ponds where the overflow water falls out. This system will work best on sloped land.

This system is just an example, you can create your own new system fitting your needs, as long as the system follows the natural patterns of your land.

6.6.7 Storing and Preserving Produce

This module has provided many ideas for growing crops. But, storing and using vegetables well is also important. Good storage means that vegetables last longer and keep more vitamins. Good storage minimize vegetables that must be thrown out and increase selling opportunities.

Almost all types of vegetables can be left in the ground until needed. However, for some types of vegetables, good storage is essential.

After crops are harvested, clean and throw out all rotten plant parts. Store in a cool place, protected from hot sun and safe from pests or other animals.

Three types of traditional containers which are good for storage:

⊙ Clay pots, for small vegetables and green leaf vegetables. Cover the top of the pot with a damp cloth and use a string or rubber to tie it on. Keep out of the sun. These vegetables will stay fresh for a few more days

⊙ In Africa, two clay pots are used, a smaller pot inside a larger pot. Damp sand is placed between these pots. Cover and keep out of sunlight. This technique works better than just using 1 pot.

⊙ Coolgardie safe. This is a simple tool made of a large box covered with wire, it uses water and wind to keep vegetables cool. This container can also be used to store meat or other foods. This container is inexpensive and easy to make. (For more information about Coolgardie safes, see Module 12 – Appropriate Technology)

If too many vegetables are picked during one harvest, and cannot all be sold or eaten, there are a few methods which can be use to store the vegetables, including:

⊙ Solar driers, which can be used to dry vegetables, fish, meat and fruits

⊙ Vegetables and fruits can be preserved as sauces, pasta, pickles and jams. Some examples: Sauces made from tomato, chili, tamarind. Pasta made from peanuts, candle nut, cashews. Pickles made from cucumber, onion, capsicum, cabbage, mango, bamboo. Jams can be made from any type of fruit, except watermelon

⊙ Some vegetables, such as eggplant, capsicum and tomatoes can be dried and stored in oil

7

FARMING

7.1 INTRODUCTION

- ⊙ Use agriculture land optimally, maximize production using minimal expenses in the mostsustainable ways

- ⊙ Form a community cooperative and farmers group which can work together; share resources, expenses, techniques and knowledge

- ⊙ Improve techniques for storing, marketing and distributing produce Working together in the community should be the main focus. These ideas can be implemented for any agricultural development, from small kitchen gardens to large community agriculture farms, on small areas of land or vast areas of land, on flat land or sloped land.

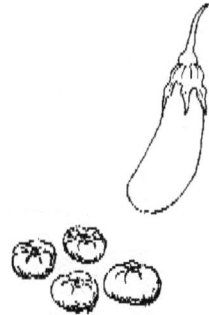

Our farmers already have a lot of knowledge and traditions related to agriculture. Therefore, this module only offers some additional knowledge and techniques to help support more sustainable agriculture.

7.2 THE LAND, ENVIRONMENT AND PEOPLE

Agriculture is a part of the land and environment around it. Wherever agriculture is practiced it will affect and be affected by the land, environment and people.

Agriculture is affected by:

- ⊙ Climate: Sun, rain, wind

- ⊙ Surrounding land and land use

- ⊙ Surrounding vegetation and animal life

- Water supply and quality
- Soil type and quality
- Erosion and landslides
- Distance from houses and towns
- Resources available to farmers and workers, such as: Seed, tools, fertilizers, harvesting equipment and so on
- Transportation and marketing of produce

If there is already an understanding about what affects us, we can choose to use simple techniques and solutions to maximize production.

Some solutions could be:

- Using terraces and swales to protect the soil, stop burning and prevent landslides
- Using natural fertilizers and pesticides, instead of using chemical fertilizers and pesticides because
- chemical materials create pollution and other problems

7.3 IMPROVING AGRICULTURE CONDITIONS

The following ideas and techniques will help you to improve crop quality, while protecting the land for future use.

7.3.1 Windbreaks

Wind is needed for agriculture and life in general. However, strong winds can cause damage to plants and trees, and many other problems for animals and people.

A windbreak can be 3-4 rows of trees planted together, which will slow down strong winds, but still allow soft winds to enter the garden. Windbreaks are very useful for flat lands and areas that have especially strong winds. Even small windbreaks can still benefit large areas of land.

Direct benefits of windbreaks for agriculture include:

⊙ Reduces plant stress, therefore increasing plant growth

⊙ Reduces plant damage caused by wind

⊙ Reduces erosion

⊙ Reduces water evaporation from plants and soil, which conserves water

⊙ Stabilizes soil temperature; the soil will not become to hot or too cold. Stable soil temperature is important for healthy plant roots and soil biota

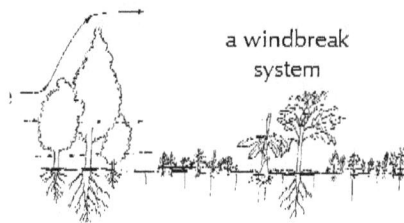

a windbreak system

7.3.2 Other benefits:

⊙ More trees will attract insects and birds, which will increase pollination rates. Increasing pollination rates will increase the amount of resulting produce

⊙ Using legume trees as windbreaks will increase the amount of nitrogen in the soil

⊙ Windbreak trees can also provide animal fodder, nuts, oils, wood, mulch, fibre, medicines and much more

⊙ Animals will be healthier because their stress will reduce

⊙ House areas will become cooler and more comfortable to live in

7.3.3 Windbreak Location

By answering the following questions, you can decide where the best locations for windbreaks are.

⊙ Which direction do strong winds come from?

⊙ Which direction does wind most often come from?

⊙ What needs protection from strong winds?

A windbreak 5 meters high will slow down winds for 100 meters of land behind the windbreak. A windbreak 10 meters high will slow down winds for 200 meters of land.

BEWARE!

⊙ Windbreak tree roots will reduce productivity of any crop grown next to them

⊙ Shade from large trees when they are fully grown will affect crops around them. Because of this, it is better not to use trees that are too tall for windbreaks

7.3.4 Constructing Windbreaks

A windbreak needs 3 or 4 rows of trees to function well. This will be thick enough to slow strong winds and direct winds to rise or fall. Use trees that will still allow soft winds to pass through, for example: Casurina, moringa, tamarind, acacia and bamboo. Trees with very thick leaves, like jackfruit, avocado and mango are not the best trees to use. Use a variety of tree types when planting. Bamboo and legumes grow quickly, and are good to use because they will become functional faster. Length and shape of the windbreak will depend on what you need it to protect. Wind will flow around the sides, so always make the windbreak longer than the area which is being protected. Windbreaks can be made:

- ◉ Zigzagged
- ◉ In straight lines
- ◉ In curved lines
- ◉ In separate sections

7.3.5 Windbreak Maintenance

Protection from animals must be provided while the windbreak trees are still young. Replant where any trees die, if possible when the trees are still small so the windbreak will grow evenly. Don't cut all the bottom branches, because the trees should be able to slow wind that might flow below the trees. Try to keep the tree shapes whole and even to achieve best results.

Precautions

- ◉ Use taller windbreaks to protect tree crops, 10-15 meters high is best
- ◉ Use some fire resistant trees. This will reduce potential fire problems

7.3.5 Swales and Water Storage

Erosion and flooding can cause serious damage, it can destroy crops, animals and even houses. Erosion can take away large amounts of soil, and the soil could enter into irrigation canals and paddies, which will cause even more problems.

Large floods or flash floods often happen in some areas of Indonesia. Some floods can be prevented, and some cannot. However, the effects of flooding can always be minimized.

Swales and other water storage techniques can be used to catch and store water, which will help to prevent erosion. This will also help to prevent stagnant water and large amounts of overflow water. Planting trees will also reduce the risk of erosion and help the swales to function better.

Make swales so that they start where water most often builds up or collects.

To achieve best results, work together on a community level, because making swales can include all lands along the water's path. (For more information about planning and constructing swales and other water storage techniques, see Module 8 – Forests, Tree Crops & Bamboo).

7.3.6 Fences

- ◉ Fences are very important for protecting your crops.
- ◉ Living fences provide mulch and animal fodder.
- ◉ Fences can also function as windbreaks.

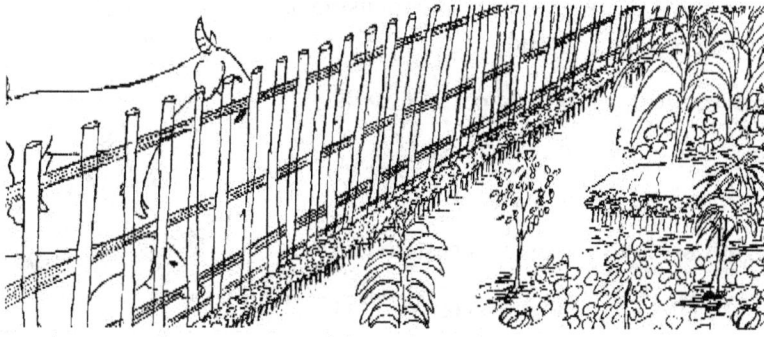

7.3.7 Stop Burning

Burning land must be stopped because:

- ◉ It causes and increases erosion
- ◉ It destroys all organic matter and soil biota which are needed to keep soil Healthy
- ◉ It kills plants which could be used as mulch material
- ◉ Fire can spread easily, especially when there is a lot of wind, which can damage crops. This happens most often during dry seasons
- ◉ It reduces bird and insect life, which function as pollinators and natural pest predators

Increase the Amount and Variety of Trees

More trees and more variety of trees growing on your land will provide more benefits. Most importantly, trees will provide protection for soil and help to prevent erosion. Trees will also attract birds and insects, which will increase pollination and function as natural pest predators.

Natural Patterns

Straight lines and squares do not exist naturally in nature. Follow the natural patterns of land. If you observe the land's natural shape, water flows, soil quality, sun direction and so on, the land will 'tell you' which shapes will work best for your land.

Terraces and swales are good examples of using natural patterns to create productive land.

Working with nature and natural patterns will:

- ⊙ Conserve energy and resources
- ⊙ Maximise land productivity
- ⊙ Improve the land's long term sustainability

7.4 IMPROVING LAND FOR AGRICULTURE

7.4.1 Organic Mulch and Fertilizers

There are many ways to provide natural fertilizers for your land.

Mulch

Use plant waste and leaves as mulch material. Some legume trees, such as leuceana, acacia, sesbania and moringa, can be planted between cropland to provide mulch. Besides providing nutrients to the soil, mulch will also improve soil because it provides organic materials and food for soil biota. By using mulch, water will stay in the soil for much longer and erosion levels will decrease.

To achieve best results, mulch must be used regularly. (For more information about mulch, see Module 4 – Healthy Soil).

Liquid Fertilizer

Liquid fertilizer is a good natural fertilizer because it is a concentrated compost. This fertilizer is very strong and must be diluted before being used. It can be used in many ways, it can be applied directly to the land or into irrigation water. To achieve best results, use liquid fertilizer before planting, during crop growth and after harvest. (For more information about how to make and use liquid fertilizer, see Module 4 – Healthy Soil).

7.4.2 EM (Effective Microorganisms)

EM is a liquid which will increase the number of microorganisms in the soil, this will improve soil quality and increase crop production. EM is good to use for agriculture because:

- It can be easily used on large areas of land

- It can be combined with any type of organic fertilizer, including liquid compost and mulch

- Microorganisms in EM will naturally and quickly multiply in the soil

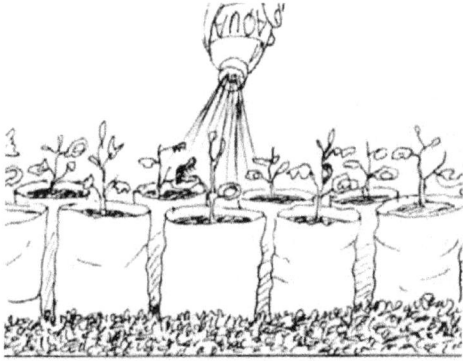

(For more information about how to make and use EM, see Module 4 – Healthy Soil).

7.4.3 Compost and Manure

Compost and manure are very good for supplying nutrients and improving soil quality. On the land, they can be used in small amounts to supplement mulch and liquid fertilizer. If animals are housed at night, it will be much easier to collect large amounts of manure. Animals can also be fenced or tethered on the land in the dry season or before planting time, so that they can directly fertilize the land.

7.4.4 Green Manure Crops

Green manure crops are only grown for soil improvement, and not as human food. These crops can be legumes, seasonal grasses and other seasonal plants. Plant green manure crops between harvest time and planting time, or on land between garden plots which is not being used.

green manure crops

crops eaten by animals

Green manure crops can also be integrated with animals. Bring animals onto the land when the green manure crops first flowers appear. The crops will provide nitrogen and some organic materials for the soil, and the animals will provide manure to fertilize the soil.

7.4.5 Ground Cover Crops

Ground cover crops are crops that grow along the ground, covering it. These crops provide nutrients and organic matter for the soil, and will help prevent weeds from growing, because the ground cover crops block out sunlight needed for weed growth.

7.4.6 Water Storage and Irrigation

For many farmers in Indonesia, crops are dependant on rain. Because of this, almost all agricultural crops are grown only during the wet season.

However, there are two important ways to increase planting seasons, by storing and protecting water in the soil and by using good irrigation techniques.

7.4.7 Storing and Protecting Water in the Soil

Swales and terraces are a good way to catch and store water in the soil. This technique is good to use, both on flat land as well as on steeply sloped land.

Mulch will protect water in the soil and stop the soil from becoming dry. Healthy, living soil will hold water better than unhealthy soil.

Irrigation

Irrigation and good water management will provide many benefits, such as:

- ◉ A continuous supply of water, even during the dry season

- ◉ Helps water storage

- ◉ Directs water to where we want it to go

Make a complete plan before implementing an irrigation system. For large areas of land, it can take a lot of time and money to make, use and manage irrigation systems well. Start small, and then think about what you want in the future. Observe where water flows through your land, and plan how to use water from one area to provide water for another area. Working together with neighbors and community groups to make and maintain an irrigation system will save time, costs and labour.

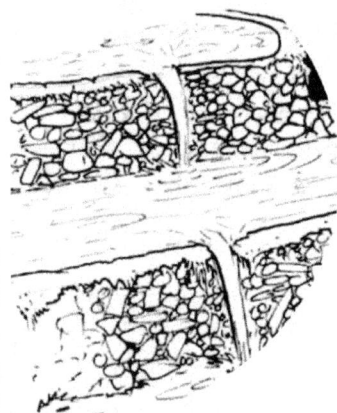

Swales can be used for irrigation. Swales can easily be combined with aquaculture systems, paddies and gardens. If you use swales for irrigation, make sure that the overflow points are well made to prevent erosion. Rocks or simple fences can also be used to manage water flow on swales.

7.4.8 Tree Swales / Terraces

Using tree swales / terraces is a good method for improving production and soil quality, while stopping erosion.

The process of making them:

- ◉ Mark out contour lines, about 5 meters apart on gentle slopes and 2-3 meters apart on steep slopes

- ◉ Dig small swales on the contour lines

- ◉ At the start of the wet season, plant legume seeds on top of the swales, about 3-5 cm apart. Legume trees like leuceana are the best type to use

The legume trees will grow into a thick fence, which will eventually create swales / terraces.

The legumes can be cut back 3-6 times during wet seasons, and 1-2 times during dry seasons. Vegetables and other crops can be grown between the rows of trees (in the empty rows). (For more information about tree swales and crops which can be planted in the empty rows, see Module 8 – Forests, Tree Crops & Bamboo).

Vetiver grass can be grown on some rows instead of legumes. The vetiver plant has very deep roots which will hold the soil together, and it will also produce lots of mulch material. Vetiver grass is especially good to use on very steep slopes.

7.4.9 Using Buffalo Ploughs

Buffaloes can be used to prepare land for planting and for ploughing contour lines. Buffaloes also provide income, meat, leather, manure and more buffaloes. Buffaloes can be expensive, but they will be able to work for a long time. Once a family, farmer's group or community has a male and female buffalo, there will eventually be a family, group and community of buffalo! To work well, buffaloes must be healthy. Buffalo maintenance requires time, but will not cost a lot of money. Food, water, shelter and medical care are all important to maintain buffalo health. Buffaloes are commonly used for ploughing land. Almost anyone can learn how to train buffalo and how to use buffalo ploughs.

The results of buffalo ploughing is good for plant growth, because buffaloes plough the land in a straight line and don't turn the soil over and over again. Turning the soil many times will damage soil structure. Some areas of Indonesia have very sloped land. Buffalo can be used on all types of land, from flat lands to very sloped lands. Buffaloes can also be kept anywhere, as long as there is water.

7.4.10 Compare buffaloes to tractors!

Using tractors takes a lot less time to prepare land for planting and can work well for large areas of flat land. However, there are many other factors that must also be taken into consideration when deciding to use a tractor or not.

Tractors are very expensive, too expensive for many people, even for farmer groups. But tractors can be rented as an alternative to buying one. Tractors cost money to maintain, including costs of fuel, oil, tyres and engine maintenance, and there needs to be someone that has a lot of knowledge about machines to operate one. Spare parts can be hard to find and sometimes need to be imported from other countries. Tractors also need a storage house to be kept in. Tractors need lots of training to operate, and some types of tractors are more difficult to operate than cars.

Almost all tractor ploughs turn the soil over and over again. This will help crops grow, but it will also damage soil structure. Therefore, soil quality will get worse over time, not better. Tractors which cut the soil in lines as opposed to turning it over are better to use. There are many areas of Indonesia with flat land, where tractors can be used. However, there are also many areas where it would not be possible to use a tractor, because the roads are too narrow or cannot be passed during the wet season. Also, many areas still do not have road access.

7.4.11 Reducing Soil Compaction

Soil compaction can cause many problems for agriculture, such as:

- The soil will hold less water
- Soil biota, which are important for soil health, will reduce in number
- The soil will contain less air
- It will be more difficult for plant roots to grow, so they will grow closer to the grounds surface
- All these problems will cause plants to grow smaller and to use more water.

The following techniques can be used to reduce soil compaction:

- Use organic mulch and fertilizers. Unhealthy soil will easily become compacted as it dries. While healthy soil which contains mulch and organic fertilizers will dry slower, and when it does dry, the soil will not be as compacted
- Use terraces and swales to shape the soil
- Use buffalo ploughs, because they will cause much less soil compaction than using tractors to plough the land
- Avoid grazing too many large animals, like buffalo and cow, on your crop land
- People can also cause soil compaction. Make garden walking paths and only use these paths when in the garden, so less areas of the garden will be stepped on

7.4.12 Intercropping

Intercropping means combining different types of crops in one area. There are many benefits provided by intercropping and there are many different combinations of crops. The type of crops used and the way they are combined is up to you.

Some combinations could be:

- Legume trees, small fruit trees, grains and vegetables: Legume trees, such as moringa and sesbania, will provide shade, mulch, nitrogen and animal fodder, and can even function as living fences and trellising which can be used to grow vine plants. Fruit trees, such as papaya, banana and citrus, can be grown together to reduce pest problems. In between these trees, vegetables and grains can be grown
- Corn and peanuts: Corn will provide shade for the peanuts and peanuts will provide nitrogen for the corn
- Cassava, small fruit trees and legume trees: With this combination, all crops will receive benefits
- Corn, pumpkins and beans: This is a traditional combination that is commonly used in many countries Mound rows with trenches dug in between is good for

holding rain water and for irrigation. The mounds can be used to plant different grains, vegetables, legume trees and small fruit trees.

The trenches can be used for planting water spinach, watercress and taro. The mounds do not have to be straight, they should follow natural land patterns.

7.4.13 Integration with Other Systems

Different systems will work better if they are integrated together as one system. Maintenance will also be less expensive, because waste from one part of the system can be used as a resource in another part of the system.

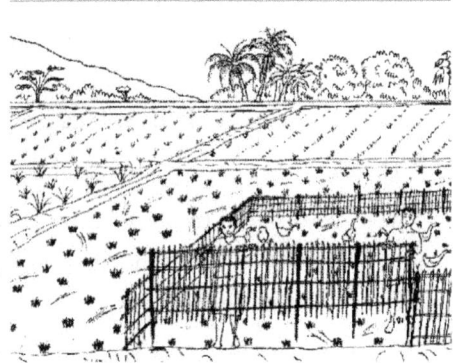

Rice paddies with ducks

Ducks can be used to clean the paddies after harvest, and at the same time they will fertilize the soil. Ducks can be easily directed from one paddy patch to another by using moveable fences.

The ducks can be kept in the paddies for a short or long periods of time. This will depend on the farmers needs, the number of ducks, the size of the rice paddy and the number of crops per year.

Trees with annual crops

Small fruit trees and legumes can be grown with grains and vegetables. The trees can be harvested to add variety to crops, besides grains and vegetables. Trees can be grown in rows or in small groups around the cropland.

Animals with crops

Animals can be grazed on croplands after harvest, they will fertilize the soil. Don't leave buffalo or cow in one spot for too long, because this could cause soil compaction. Legume trees planted in rows or to divide cropland into sections, can be used as living fences for animal grazing.

Rice paddies with fish (mina padi)

With careful management, fish can be kept in rice paddy water channels and in the rice paddies at certain times of the year. (For more information about this technique, see Module 11 – Aquaculture).

Ponds with croplands

Water from aquaculture ponds is nutrient rich and should not be wasted. Swales, terraces and paddies can be used to catch and store overflow water, and this water can then be used to fertilize vegetables and trees.

7.4.15 Natural Pest Management

It is best to prevent pest problems before they occur. A healthy system, with healthy, nutrient rich soil, will experience far fewer pest and disease problems. There are many natural pest predators and natural pesticides which can be used if problems do occur.

7.4.16 Weed Management

Weeds are an issue which strongly affects agriculture. If weeds are not controlled, crop production will decrease. Weeds use nutrients and water from the soil, so they are competing with the main crops. However, if weeds are used as mulch, some nutrients and water will be returned to the soil. Don't burn weeds, because if weeds burn their benefits will also burn.

Some weed management techniques:

- Use mulch to cover the soil, the thicker the mulch layer the less weeds will grow

- Grow ground cover crops, like pumpkin or beans, to block out sunlight which weeds need to grow

- Ploughing the ground before planting will turn most weeds into the soil

- Clear away annual weeds

- Use animal labour to clear weeds, animals will receive food from weeds and the soil will receive manure from the animals

- Remove weeds before weed seeds form, this technique works well for perennial weeds

⊙ Control irrigation, the more directed water is, the less weeds will receive water. Pipe irrigation is the best, because only main crops will receive water

7.5 SYSTEM OF RICE INTENSIFICATION (SRI)

⊙ SRI is a method for increasing rice production. This method is already being used in many countries, including Indonesia. Using this method, rice crop yields can double, in comparison to other methods.

⊙ Besides increasing production, SRI will also provide many other benefits, such as less water usage, saves seeds, is environmentally friendly, and it needs far fewer external inputs.

⊙ SRI can be used on large or small rice fields, it does not need new machinery, tools or special fertilizers. This method has been successful both with traditional seed varieties and non-traditional seed varieties. This method can be used in areas with limited water supply. Because of this, it will also extend planting seasons. SRI works best when combined with organic fertilizers and natural pest management techniques. So that SRI methods work well, training and practice is needed in the beginning, until farmers increase their skills. Good irrigation methods and water control are also very important.

SRI Techniques

1. Early Transplanting

Plant seedlings when they have just two leaves and the seed sac is still attached, this is usually 8-12 days old, sometimes up to 15 days, and in colder areas could even be 16-18 days.

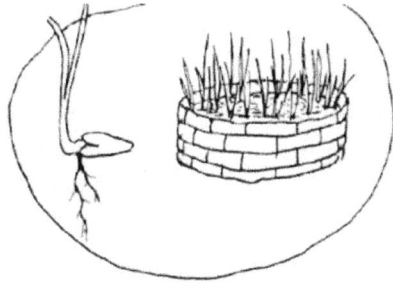

Early transplanting gives the rice the maximum time needed to root, leaf and grow. Every day delayed reduces the growth potential, especially after 15 days.

2. Careful Transplanting

Plant the seedlings in muddy soil, not in standing water, with the roots about 1-2 cm deep and the root tips pointing down or across. If the seedlings are pushed into the soil, their root tips will point upwards and this is not good because the seedlings growth will be slowed or stopped for up to one week as the plant recovers.

Careful transplanting will reduce root damage and plant stress, while reducing delays in plant growth after transplanting. This will have a big impact on the plants growth later on.

3. Plant Spacing

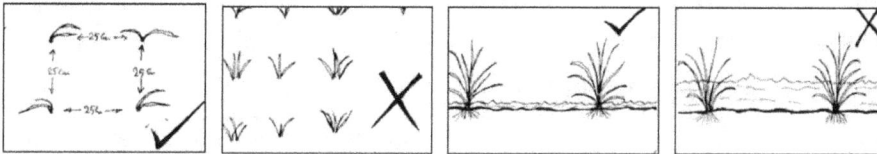

Plant the seedlings one by one, not two, tree or four at the same time. Seedlings are planted in square patterns about 25 cm x 25 cm in size. Plant spacing can be estimated or marked out using anything which will measure well, such a special rake to define the planting points. Using this side spacing will promote better root and leaf growth.

4. Watering and Well Drained Soil

While the plant leaves are growing, only give enough water to keep the soil moist, make sure not to water too much till the water is continuously stagnant. When the rice plants start to flower and form grain, maintain a thin layer of water for all the plants of about 1-2 cm. As usual, drain the water before harvesting. Well drained soil will promote much larger root systems.

5. Early and Frequent Weeding

Start weeding 10-12 days after planting seedlings, use a simple rake or hoe. Weed ever 10-12 days following, until the rice grows large enough to shade all of the ground (forms a canopy). In experiments, every weeding increases the rice yields per hectare till up to one ton! Frequent weeding adds air to the soil which improves root growth, and also removes the weeds which are competition. Mulch can also be used to prevent weeds from growing.

6. Apply Compost

SRI works well without compost or fertilizers, but using natural compost and fertilizers will improve plant growth, improve soil quality and increase harvest yields. Experiments have shown that organic composts and fertilizers provide better results compared to chemical fertilizers, especially over longer time periods. This is because of improved soil quality and microbe activity in the soil, which increases the amount nutrients in the soil which are available for plant use. Mulch is also very important for providing nutrients and increasing soil biota. Using EM (Effective Microorganisms) will also help to improve results.

These techniques give good results for plant growth because healthy root growth leads to healthier stalks and leaves, larger rice grains, and stronger, larger seedlings.

Using SRI

Farmers and groups interested in SRI should have enough training in SRI methods to increase their skills and knowledge before practicing SRI in their fields. The training could be small experiments on 1 or 2 rice paddies to test the results and compare them with techniques already being used.

7.6 WORKING TOGETHER

Agriculture is dependent on the environment around it and it affects the environment around it. Therefore, it is important for people and communities to work together in agriculture practices, such as sharing resources like water, labour and tools. Working together can start from families and community groups, which can lead to national or even international levels!

7.6.1 Community Consultation

Just a few individuals can possess a wealth of knowledge about agriculture. Sharing and gathering information will help improve agriculture results for everyone. Everyone involved in agriculture production should be part of this process, because most knowledge comes from observation and practice. Women should be involved and need to be included in community consultations because they are involved in many of the daily agriculture practices, and hence have a lot of observation and knowledge to offer.

This knowledge and information could be:

⦿ What is the best age and time to plant seedlings?

⦿ What pests are attacking crops?

⦿ Are there any natural predators eating the pest?

⦿ Are there different areas on the land where crops grow better than on other areas? Why?

This important information and more is needed to improve land and crop management.

7.6.2 Community Participation and Understanding

The more people in a community that understand and participate in agriculture development, the more productive and sustainable agriculture practices will become. Issues which require community participation and understanding include water management, crop harvesting and marketing, use of chemicals, waste management, fences and much more. For example, in water management alone the following issues may need consideration:

⦿ What affects water sources and how does it affect the water source?

For example, if the water source is a river or spring, land above it affects the water source, and the water source will affect the land below

⦿ How to protect the water source?

⦿ How to collect water? For example, from rivers, making wells or water pumps

⦿ How to direct water? For example, using pipes or water trenches

⦿ How to divide water between all the people who use it

⦿ Who will pay the costs for building irrigation systems?

⦿ How will irrigation systems be maintained and repaired?

⦿ If farmers and community groups know and understand these issues, it will be possible to find solutions together, and then the benefits received will also be enjoyed together.

Benefits could be:

⦿ Water is shared in the best way

⦿ Costs are reduced

⦿ Management and maintenance becomes easier

⦿ All factors up river and down river are considered together

⊙ The environment and water quality can be protected and improved together

7.6.3 Working with Neighbors

If people work together in a community, the whole community will benefit.

Working together with neighbors is important and will give benefits for everyone involved.

Avoid all forms of competition and jealousy! This is useful and beneficial for the future.

7.6.4 Community Cooperatives or Farmers Groups

By forming a cooperative, the following benefits will be achieved:

⊙ Resources can be bought for less personal expense

⊙ Harvesting, storing and marketing produce will become easier and more efficient

⊙ There will be a place for sharing knowledge, labour, seeds, tools and other agriculture products

⊙ All community assets will be protected and conserved

⊙ It will be easier to be heard by other groups because the government and large organizations are more likely to listen to a group of people than an individual

(For more information about community cooperations or farmer groups, see Module 13 – Cooperatives and Enterprise Development).

7.6.5 Working with Nature

Agriculture practices which work in harmony with nature will give better results, especially in the long term. There are many techniques which can be used to work with nature, and not against nature. One example of working with nature is to plant crops during the wet season. It is best plant crops when the rain has already come 3-4 times, at the beginning of the wet season. Planting before this often causes crops to grow smaller because there can be breaks between one rain till the next rain.

Perhaps planting at the right time causes some difficulties, such as food shortages or labour management. However, planting at the right time will achieve maximum crop yields.

7.7 POST HARVEST STORAGE AND USE

Post harvest is a time when there is a lot of food crops and there is also a high potential for these crops to be wasted or lost. This could mean large income losses. There are some things which can be done to reduce these losses.

For example, for beans and other dry grain produce, make sure to:

- ◉ Harvest at the right time

- ◉ Separate seeds from plant materials as soon as possible, this will reduce insect problems

- ◉ Dry produce properly, because if produce is stored when not dried completely, it could cause rotting

- ◉ Store produce properly. Use dry, secure containers, and protect them from insects and mice

Natural Protection from Insects

Insects can be a problem when storing harvest crops. Natural material can be used to protect crops from insects.

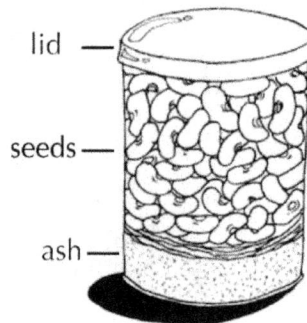

Some of these natural materials include:

- ◉ Kitchen ash. For storing large amounts of grains, add 2% of ash to the weight of the grains to be stored (for example, for 100 kg of grain add 2 kg of ash). For small containers, add a 1 cm layer of ash at the top and bottom of the container. Don't use ash from rubbish fires!

- ◉ Tobacco leaf. Use old, dry tobacco leaves, and only use for large storage containers. Add a 2 cm layer of tobacco leaves on top of the grains in the container. Be careful using tobacco leaves because they are very strong

- Gamal leaf. Add a layer of about 2 cm of dried gamal leaves on top of grains to be stored

- Neem leaf. These leaves can used fresh or dried. For large containers, add a layer of neem leaves about 2 cm thick on top of the grains to be stored, for smaller containers a layer of 1 cm will be enough

- Fruit peels. Lemon, lime, grapefruit or orange peels will effectively repel insects from stored produce

- Eucalypt leaf. Use 10-20 fresh or dried and crushed eucalypt leaves. Spread them over the grains that will be stored

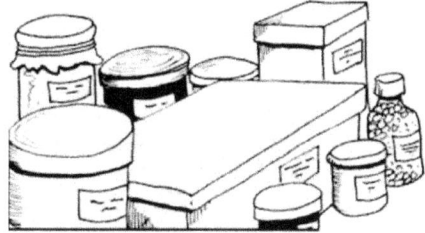

For fresh produce, like tomato, lettuce, beans and cabbages:

- Harvest at the right time

- Produce which must stay fresh should be kept somewhere cool, dry and protected from insects and animals, until it will be eaten or sold. Spray vegetables with water to keep them moist

- Transport the produce carefully, keep in a cool place, away from sunlight and avoid bruising as much as possible. This will make food last much longer

- Store in a clay pot with a damp cloth on top. The produce inside will stay good for many more days

For root vegetables, like potato, cassava, taro and yams:

- It is best if root vegetables are left in the ground until they will be eaten

- For selling, harvest as carefully as possible. Cuts, bruises or any damage will cause these roots to rot faster

- After harvesting, store root vegetables in a cool, dry place, out of the sun and protected from insects and animals. Wood ash can also be used to protect from insects and animals, but make sure to wash them before selling

- Carrots are different, because if left in the ground for too long they will become hard and bitter. Store them in sand to make them last longer.

Using Excess Produce

Sometimes not all the produce can be eaten, sold or traded, and some will go rotten. But this produce does not have to be wasted or thrown out.

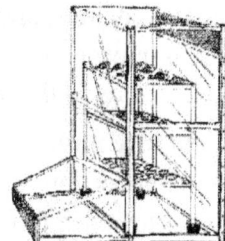

Here are some simple solutions:

- ⊙ Use solar driers to dry vegetables or fruits, so that they can be stored and used later on. (For more information about how to make and use solar driers, see Module 12 – Appropriate Technology)

- ⊙ Make sauces, preserves, jams and more from vegetables and fruits

- ⊙ Excess food can also be used as pig or chicken feed, or turned into compost

7.8 HEALTHY AGRICULTURE

Agriculture changes the environment around it. It is very important to be aware of changes which occur in the soil, water and people. Therefore, you can prevent negative impacts on the environment, and make the land more sustainable for you and future generations.

7.8.1 Protect the Surrounding Environment

An important part of agriculture is to protect the environment around it. Small and large rivers especially need protection. Clean water and healthy rivers are essential for our future.

A healthy environment will also help to improve agriculture land. Cleaner water will reduce irrigation maintenance needs. A healthy environment will also bring more animals which are important for people and will add beauty to the environment.

7.8.2 Prevent Cropland Soil Erosion

Soil erosion from croplands will deplete your croplands. Erosion can also cause huge problems for lands below. These problems will continue on to rivers, and eventually to the ocean.

Soil erosion can be prevented by:

- ⊙ **Catching and storing water.** Use swales and terraces for sloped land, even on only gentle slopes. For flat lands, it is important to control water flows on higher lands so that the water doesn't build up on the cropland

- ⊙ **Use mulch and stop burning.** Burning damages soil structure, which makes it easier for soil to erode. It also destroys plants, which function to hold the soil together. Mulch protects the soil and improves soil quality, so that produce yields will increase each year Croplands and areas around which have already eroded need to be replanted. Grasses, bamboo and fast growing legumes are best to plant on eroded lands.

8

FORESTS, TREE CROPS & BAMBOO

8.1 INTRODUCTION: AN OVERVIEW

- People have a strong and continuous connection to the land and forest. Forests provide food, wood, natural materials, medicines, fuel, homes for animals and birds, and a spirit connection with ancestors or animals which lived there long ago. Areas that have forests need to be protected and carefully managed. These forests are the seed banks of the future.

- We have many plants and animals which can only grow and live in our environment. This is because of climate, landscape and the way the land was formed long ago. Preservation of these species will help us to retain our culture and heritage. Many of these plants are medicines, contain oil and other useful products which can bring income in the future.

- The first step is to protect and carefully manage the forests. The next is to reforest and restore natures balance. We must maintain a strong connection with nature. We need long term solutions for keeping the environment and land healthy and strong for the future.

- Many areas where forests have been removed are suffering from erosion and soil loss. It is difficult to obtain good productivity on these lands. In fact, agriculture practices on these lands can even create more erosion and new problems.

- Reforestation and tree crops can help stop erosion, repair damaged land, while providing food, wood, oils, medicines, fibre and many other products for income. These are all sustainable incomes.

- Tree crops can also be integrated with animals and annual crops. Products and income from trees and forests are more secure because trees are less affected by bad weather conditions.

⊙ A well designed forest system will need little maintenance once it is established. Forests and trees will improve the health of the environment, not just on the land where they grow, but also on the land surrounding it. Healthy environmental improvements in the mountains will even affect environments on the coast and in the ocean.

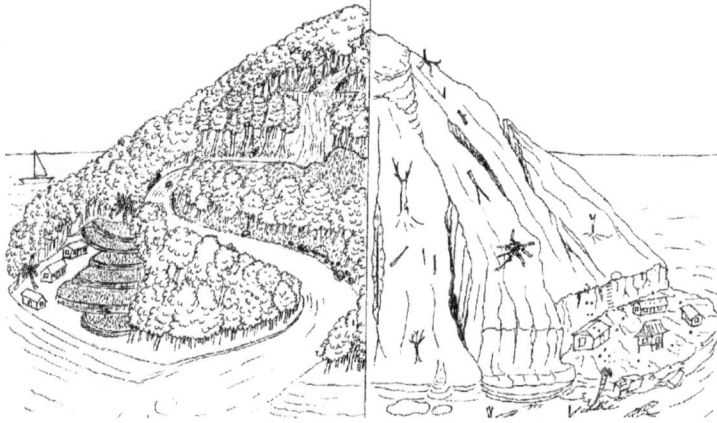

8.2 SUSTAINABLE FOREST SYSTEMS

Steps to achieving sustainable forest systems:

⊙ Store water in the ground

⊙ Protect soil and stop erosion

⊙ Control animals

⊙ Stop burning

⊙ Forest and resource management

8.2.1. Store Water in the Ground

Water is precious! Water stored in the ground will benefit the land, plants and people.

Some of these benefits are:

⊙ The soil will be protected from erosion

⊙ Crop seasons will be longer

- Plants will be more fertile and can grow even during the dry season
- Ground water levels will rise so springs won't dry up

8.2.2 Protect Soil and Stop Erosion

Erosion will reduce productivity by removing a very valuable layer of soil. Soil, especially soil which is good for agriculture, takes a long time to form, but can be lost very easily and quickly due to erosion.

If not controlled, erosion will quickly get worse and create bigger problems in the future.

Erosion will also destroy all small plants, seeds and organic matter. Erosion on cleared land can cause landslides, which not only destroy land, but can be very dangerous for people.

future rivers without reforestation, erosion or animal control

future rivers with reforestation, erosion and animal control

Reforestation and tree crops are a long term solution for protecting the soil and stopping erosion. Making swales and terraces will help hold water, this is an important base for reforestation, tree crops and all sloped land agriculture.

8.2.3 Control Animals

Animals, especially goats, can damage reforestation and tree crops quickly.

This can waste a lot of hard work and useful tree crops.

Some solutions which can be used are:

- Village regulations can be made to protect specific areas

- Make fences especially for reforestation areas and tree crops
- Make small fences (tree guards) around each individual tree
- Tie up animals
- Work together with the community to control animals

8.2.4. Stop Burning

Burning land is not recommended because it can:

- Cause erosion problems
- Reduce plant and animal diversity
- Destroy natural mulch, soil biota and other organic matter important for the soil
- Cause water loss
- Pollute the environment
- Damage reforestation and newly planted tree crops
- Reduce certain resources

Many areas are burned to encourage new grass growth for animals. This is achieved, but it causes the land to become unproductive in the future.

Also, burning will only encourage low quality grasses to grow for animal food.

Burning should be stopped because the damages caused are larger than benefits received.

Think of better, alternative solutions to replace burning land.

8.2.5. Forest and Resource Management

The following are important steps for forest and resource management:

- Make a community management plan
- Replant trees which are cut down for use

- Make village regulations for protection community forests or certain areas which need protection

A community management plan is a plan for the future.

A community management plan should:

- Plan what can be harvested and whom can harvest it

- Control income from harvests, including how much income should return to the community to be used for forest management

- Develop a working relationship with the government

- Protect community land from other groups interested in ownership, for example from commercial enterprises

- Use forest resources wisely. Beneficial forest resources include seeds, medicines, oils, bamboo products, honey and much more

Smaller community groups, like women's groups and youth groups, can compile information and important guidelines regarding land and forest management. This information can be passed on and discussed at community meetings, so that knowledge and ideas can be included in community management plans.

Collecting firewood can cause huge losses to now existing trees. Collecting firewood is also hard work, which takes up a lot of time. Finding alternatives to firewood is important, because by reducing the use of firewood a lot of time and energy will be saved, while protecting the forest. Take into consideration changes such as:

- Planting trees around the house which can be used as firewood, this could even be living fences

- Using stoves and ovens which use less firewood or none at all, like charcoal stoves

BEWARE!

International companies are looking to our forests to make money for themselves. This could be from harvesting forest resources or clear cutting for commercial purposes, such as for farming land or other reasons. Short term jobs and small amounts of money will never be able to replace the wealth and value contained in forests. Commercial companies will always take most of the money and benefits, this is how they work. This is already happening in many countries all over the world and it is causing huge problems and destruction.

Any plantations should be separate from the forest area and forests should never be replaced by plantations. Our beautiful forests and environment can be an important asset in the future, such as for ecotourism, which can provide many more jobs and local income than logging and plantations can. This is also more sustainable for the future.

Making Swales

A swale is a trench that is dug on contour with the land, at an equal level above sea level along a slope. The soil and rocks dug from the trench are put just below the trench to form a long mound. This forms

a level line from one end of the swale to the other end. Swales can also be small walls built from rocks, branches or other materials. Usually swales are dug on a hill side, one below the next.

Terraces are similar to swales, but swales are better at stopping erosion, catching and storing water, soil and mulch, and they function faster.

Swales will help to improve soil, catch water and stop erosion. Swales create new micro climates on the land, which means they provide new areas for planting all kinds of plants. Swales can be used for different combinations as well, like combining vegetables, trees and animals, so more variety of produce can be harvested. Swales provide long term productive solutions for sloped lands, and can be used for small or large amounts of land.

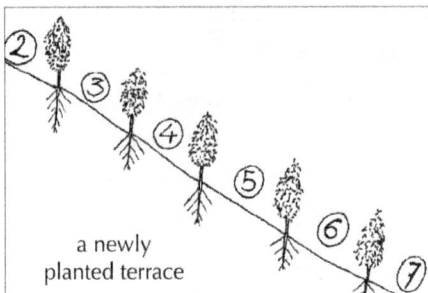

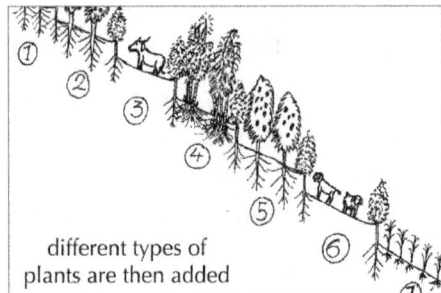

a newly planted terrace

different types of plants are then added

Swale Sizes

On gentle slopes, swales need to be wider in size, about 3-10 m depending on the situation. However, on steep slopes, swales are much smaller in size, about 1-2 m, because water flows faster. Size also depends on what you want to plant. For vegetables swales can be made closer together, and for trees further apart.

If the land is extremely steep, water will flow very fast. Because of this, it is better not to dig swales because the water will destroy the newly built swales. The best way to make swales on steep land is by planting fast growing trees on contour, the same contour as used to dig swales. The trees need to be planted close together so they will grow into a living swale, which will slow water and reduce erosion.

Rocks, tree branches or other materials can be placed against the living swales to help hold more soil and water. Eventually the soil will build up behind the trees and form a small terrace.

Vine plants, like pumpkin, luffa and passion fruit can be planted on the newly formed terrace.

Marking a Contour Line

Swales must be made on contour, if not water will flow into the swales and damage the lower part of the swale, especially during the wet season. If swales are on contour, water will be held in the swale and will soak into the soil evenly.

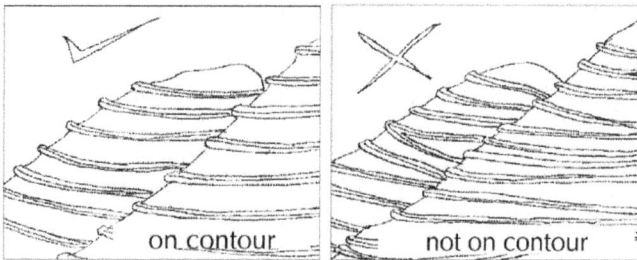

on contour

not on contour

Making an A-Frame

An A-Frame is a measuring tool made of wood or bamboo, and it is used to help make swales which are on contour. This tool is about 2 m high and shaped like an 'A'. A-Frames are very easy to make and to use.

Materials needed:

- ◉ 2 equal lengths of bamboo or wood, about 2 m long
- ◉ 1 length of bamboo or wood for the crosspiece, about 1 m long

- ⊙ 2 m string or rope
- ⊙ Hammer and cutting tool
- ⊙ Rope or binding wire
- ⊙ 1 small rock
- ⊙ Pencil or marker

Constructing an A-Frame:

Shape an 'A' from the bamboo or wooden sticks. Make sure that the crosspiece is the same distance from the top on both sides.

Tie string or rope to the top of the A-Frame and tie the rock to the bottom of the string. The rock must be positioned just below the crosspiece.

Stand the A-Frame on flat ground and mark the ground where the legs are positioned.

Mark the crosspiece with a pencil or marker exactly where the string touches it.

Turn the A-Frame around and place the legs on the markings already made on the ground.

Again, mark the crosspiece exactly where the string touches it.

The A-Frame is exactly level when the string is in the middle of these two marks. Mark the

Middle.

Using an A-Frame

Step 1:

Observe the area where swales are to be made. Decide how many swales will be made and which areas of the land will be used. Remember to use the ideas about swale size and distance between swales.

Step 2:

Start at the top swale. Cut down tall grasses or weeds which may obstruct contour line markings.

Step 3:

Place the A-Frame on the ground, position it so that the string touches the middle marking on the crosspiece. The A-Frame is now on contour. Place two stakes where the A-Frames legs mark to form the beginning of a contour line.

Step 4:

Swing the A-Frame around with one leg still in place where a stake is marking. Then, repeat step 3. Eventually a line will form along the hillside, this is the contour line.

Step 5:

Start on the next line. Continue until all the contour lines needed are marked out with stakes.

SMART IDEAS!

- ◉ When using an A-Frame, it is a lot easier and faster with two people. One person can operate the A-Frame, while the other can mark out the contour line with stakes

- ◉ Don't place either legs of the A-Frame on rocks, mounds or in holes, because this can make the line inaccurate, which could cause problems later on

Constructing Swales

There are 3 main types of swales:

- ◉ Trench Swales. Swales are called trench swales if trenches are dug, then soil and rocks that are dug up are placed below the trench to form a mound.

- ⊙ Ploughed Contour Lines are when a line is ploughed along the marked contour line

- ⊙ Rock swales are made from rock mounds or walls, without digging. Usually rock swales are made where:

  - The land is too hard to dig

  - There are many rocks available

  - There is a very steep slope

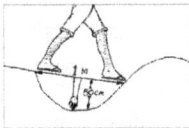

Which type of swale you use is up to you.

If needed, you can use all three types of swales on the same land. All types of swales should be planted with thick, fast growing legumes as soon as possible, to:

- ⊙ Hold the soil

- ⊙ Provide tree terraces

- ⊙ Provide fertilizer and nitrogen

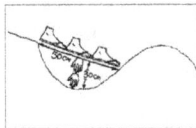

| on very steep slopes | on gentle slopes | to help control erosion, wood and rocks can be added |

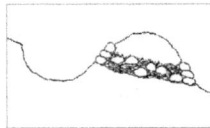

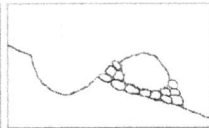

Trench Swales

Always start making the swales from the top. Dig the trench above the contour line, then make an even mound below the trench. The trench size depends on the slope of the land. On very steep slopes, trenches can be made about ½ meter wide, and about 30 cm deep.

On gentle slopes, trenches can be made about 1 meter wide, and 40-50 cm deep. Continue until all the trenches are dug and the swales are formed.

For best results, make all the swales about the same size and shape. This will help water flow into the swale, and not along the swale. This can be easily tested. Flow water into the swale when it is almost finished and observe where the water flows. Make changes if necessary to make the bottom even. If there is no water available, wait until the first rain to observe water flows. The more accurate in size, the better.

On very steep slopes on gentle slopes to help control erosion, wood and rocks can be added

SMART IDEAS!

The higher the mounds, the better. To make the mounds higher, first place rocks and dry branches, and then put the soil on top.

To protect the soil mounds from erosion, add a thick layer of mulch. It is good to mulch the trenches as well. Plant the mounds as soon as possible, use vegetables, trees or vine plants. The swales will catch and store rain water, especially in the wet season, which will help supply water in the dry season. However, plan for what could happen in extreme conditions, like what could happen in very heavy rains, if swale water overflows or if the slope is steep and water flows out of control during the wet season.

Problems caused by extreme conditions are minimized if the swales are made on contour, this will also prevent water from collecting in one spot. Water collecting in one spot could break the swale.

In preparation for overflow water, make one end of each mound lower. This will make water flow in the direction you want it to if the water reaches a certain height. Place rocks around the mounds overflow points to prevent erosion. A hole dug near the overflow point will catch soil from the water before it flows out. This will help even more to reduce erosion, the soil which is caught is good quality, it can be dug out and reused. The overflow water from one swale can be run into the next swale, which can then run into the next swale, and so on. Eventually, water can be run into aquaculture ponds or water storage ponds.

Ploughed Contour Lines

This is a very simple method, it is just ploughing along the contour lines. Ploughing can be done using a buffalo plough, hand tractor or anything else that will work. The ploughed line should be made just before the wet season, so at the beginning of the wet season it will be ready for planting.

Ploughing contour lines takes only a small amount of time, so it is good to use this method for large areas of land. However, it will take much longer to show results, sometimes up to 1-2 years longer.

Rock Swales

Rock swales are good to use on land with many rocks or where the soil is too hard to dig. Rock swales can be used on large areas of land, steep sloped land, or very small, tight areas of land. The way to make rock swales is very simple. After planning how many swales you want to make, just build rock mounds or small rock walls along the contour line to form the swales. These rock swales can be knee high on an adult, or if possible even waist high.

if possible even waist high.

making rock swales

rock swales create natural terraces

Eventually, soil will wash down the slope with rain and be stopped by the rock swales. This process will create terraces. After the terraces begin to form, the swales can be slowly raised higher. Plant legume trees as soon as possible on the newly formed terraces. The trees will improve the soil, provide fertilizer and mulch, as well as provide shade for other crops later on. In the future if you need more space for other plants, some of the legumes can be removed.

SMART IDEAS!

- Place larger rocks at the back of the swale (on the bottom side) and smaller rocks at the front of the swale (on the top side). Smaller rocks will hold more soil and water than larger ones

⊙ Rock swales are natural fences for animals, like buffalo and cow, to control the areas where they eat. If necessary, gates can be made so that animals can pass through. This can also be used for goats, but goats might climb over the rocks, so other forms of fencing will also be needed

Planting Small Swales

During the Wet Season

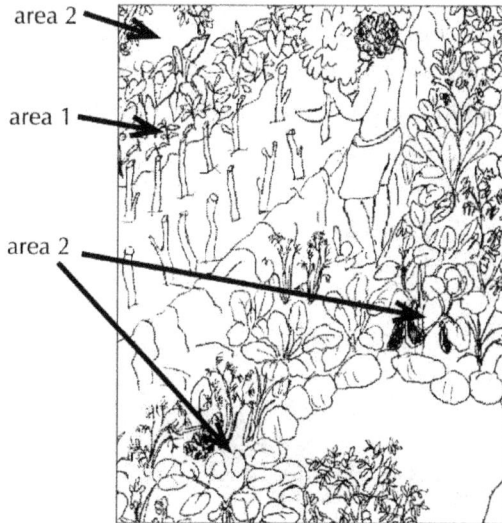

Area 1: Taro, kangkung, watercress and other water plants can be planted along the bottom of the trenches. Plants that like water, but don't like being under water, can be planted along the edge of the trenches, for example lemon grass.

Area 2: Other vegetables and small plants can be grown on top and below the mounds, such as tomato, eggplant, pumpkin, cassava, capsicum, corn and many more types of vegetables.

Rock swales create natural terraces

During the Dry Season

If there is enough water available, you can continue to use the same planting ideas as for during the wet season. If only a small amount of water is available, the trenches can be used to plant vegetables, and the mounds can be used for planting long term plants, like cassava, banana and eggplant.

If there is no water available, mulch the land and wait until the next wet season. Some long term crops can still be grown, and will grow a lot better than without swales, because swales still hold some a small amount of water in the soil. Small animals can also be integrated into this system. Remember to use soil improvement techniques to achieve the best results from swale gardens.

SMART IDEAS!

Swale trenches can also be used as compost trenches. (For more information about compost, see Module 4 – Healthy soil).

Planting Large Swales

On large swales many different crops can be planted, from perennial trees to annual vegetables. Animals can also be integrated.

The more diversity of crops and animals you have the better. The types of crops you choose is up to you, as long as those crops can grow well in your area and will provide sources of income, such as food, wood, oil and other resources.

BEWARE!

Introducing new types of plants, especially from overseas, could become a problem in the future. First, find out:

- If the new plant could become a weed and compete with local plants?
- If the new plant can introduce new pests or disease?
- If the new plant has caused problems before in other countries?

This is important for protecting our environment and resources for the future.

Agriculture Systems on Swales

At the start of the wet season, plant legumes along the swale mounds or along the ploughed contour lines. Plant the legumes close together, about 3-5 cm apart. Leucaena and moringa are both good legumes to use.

These legumes have many functions, they will:

- Hold the soil together and eventually form a fence which can be continuously pruned for many years. The pruning can be used as mulch and fertilizer, and as the legumes are pruned they will release nitrogen into the soil
- Act as a windbreak, which will help to protect crops when they are still young
- Eventually form natural terraces on the land

In between the legume rows, there is wide rows or 'alleys' which can be used for planting many different types of crops or even for grazing animals. Not all of the land must be used straight away, it is better to utilize the land gradually as needed.

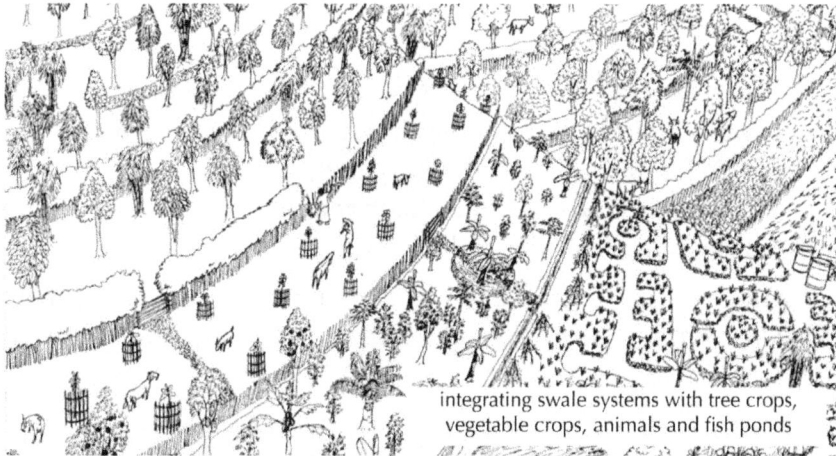

integrating swale systems with tree crops, vegetable crops, animals and fish ponds

Managing Planting Times

Managing planting times is a technique which can increase crop yields by working with crops of different sizes, different growth rates and different life spans.

Following is an example of managing crop planting times:

- **Year 1:** Plant legume trees, like leuceana and moring. Leave space for fruit trees. Legume trees will grow quickly and can later be cut back to provide more space for other trees

- **Year 1 and 2:** Plant fruit trees, like apples, mango and citrus, between the legume trees. When planting, think about how large the tree will be in 10-20 years ahead and leave enough room for trees to grow to their full size. The legume trees will provide some shade for the fruit trees when they are still young. When the fruit trees grow larger, the legumes can be cut back to provide more space. Eventually, the fruit trees will take over the legume trees. Animals can also be integrated into this system

- **Year 1-5:** There will be space between the fruit trees for about 5 years. The space can be used to grow vegetables like corn, pumpkin, beans, sweet potato, capsicum, taro, cassava, papaya, banana, pineapple and root plants, like ginger. To allow sunlight in, prune back legumes which grow too thick, the prunings can be used as mulch material

- **Year 5-10:** There may be some space still available for growing vegetables and small trees. However, these smaller crops will need to be removed once the larger fruit trees have grown. Continue cutting back the legume trees, and if more space is still needed, the legumes can be removed The more variety of crops, trees and animals there are, the more variety of foods and products there will be, and this will assure a more stable income. For soil with many rocks, dry areas or large amounts of land, plant more trees than vegetable crops. Trees require less maintenance and will still produce crops in harsh conditions.

Planting in Swale Alleys

Some examples of crop combinations for swale alleys are:

- Small fruit trees grown together, like citrus, mango, cacao, guava, apple, papaya, sesbania, coffee and taro. Spices, like ginger, chillies, cloves and tumeric, and vegetable crops, like sweet potato, spinach and cassava, can be planted all together between the fruit trees in the swale alleys

- Large trees grown together, like mango, avocado, jack fruit, coconut and bamboo. When these trees are established, in about 4-5 years, animals can be grazed between them in the swale alleys

- Timber trees, oil trees, bamboo, fibre trees, medicinal plants, firewood trees and other crops can also be grown together. Eventually, animals can be grazed in the alleys. Short term crops, like spices, sweet potato, pumpkin, papaya and even banana, can be grown when the larger trees are still young

- Smaller swales can be made in the alleys to increase production and crop variety. Use any crop combinations you like!

SMART IDEAS!

Planting many crops together and rotating different crops will help to keep the soil healthy.

Integrating swale systems with tree crops, vegetable crops, animals and fish ponds

8.3 FLAT LAND TREE CROPS

Flat land is generally used for grains, vegetables and paddies, but tree crops can be combined in many ways. Tree crops will increase production and crop variety. Tree crops need less maintenance, and will still produce in the dry season. Small trees, like citrus, banana, papaya, clove and pigeon peas can be planted with grains and vegetables. The trees will provide shade for smaller annual crops. They will also act as a barrier and make it harder for pest insects to pass from one plant to the next. Also plant legumes, they will provide many benefits.

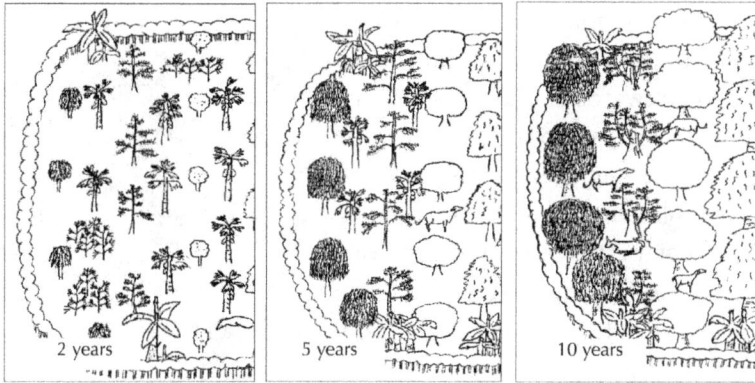

Another benefit from planting trees with crops and vegetables, is that the smaller crops can be harvested first, while waiting for the large trees to grow and produce.

SMART IDEAS!

Flat land agriculture will be improved if water is collected and stored in the ground, this includes rain water and water that flows down from the mountains. Continue using trenches and compost pits to collect water.

8.4 REFORESTATION

Reforestation areas are areas where the natural forest is restored.

- ◉ Reforestation is a less intensive system, and will provide less produce than agriculture.

- ◉ However, this system is very important for preserving the environment and stopping erosion, and it will provide many essential products, such as bamboo, oils, fibre, timber, honey and medicines.

8.4.1 Dry Land Strategies

In dry areas, water storage is very important. For dry, rocky areas, rock swales can be used. There are also other techniques which can be used, like boomerang swales and "net and pan" systems.

8.4.2 Boomerang Swales

Boomerang swales are named after and share the same shape as the traditional hunting weapon of the Australian Aboriginal people. Boomerang swales should be a minimum of 2 meters long, but will work better if they are between 5-10 meters long. They are usually about adult knee height, but higher is better. The swales are made of rock mounds, a combination of dug swales and mounds can also be

used, as long as it will still hold water. Put smaller rocks on the front side (top) and larger rocks on the back side (bottom), just like when making rock swales. This will help to collect more water, soil, leaves and plant materials.

Trees will help to hold and improve the soil. Start by planting trees in the middle of the swales, and move outwards as the trees become established. Some good trees to start with include legumes, and using the 'seed ball' technique which will be explained following. If many boomerang swales are made together, excess water from one swale will be collected in the next swale. If the system is managed well, this will increase production for all the swales.

8.4.3 "Net and Pan" Swales

These swales are similar to boomerang swales, except that theyhave on side shaped as a 'V'. This system is called "net and pan", beause a "net" is shaped to catch water and a "pan" is shaped to hold water. This system works best on gently sloped land. Each side of the 'V' shape is about 3 meters long and about adult knee height. The swales can be made of rocks or mounds of dug soil, or a combination of both.

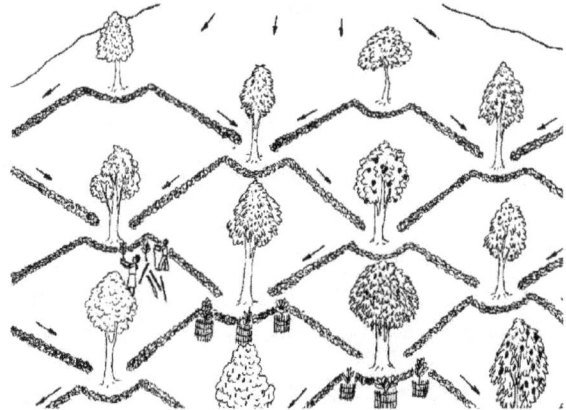

If many swales are made in an organized way, they will make a system where the overflow water from one "net and pan" will flow into the next "net and pan", and so on. Use small trenches to help direct the water.

The benefits of this system are many, but primarily it will help to reduce and prevent erosion.

8.5 MICRO CLIMATES

A micro climate is the climate of a particular area. This area could be as small as a garden plot or as large as a mountainside. Each type of plant prefers different micro climates. However, almost all plants like micro climates with:

⊙ Available water

⊙ Good soil

⊙ Enough sunlight

⊙ Protection from strong winds

⊙ Shade, for when plants are still young

Providing a good micro climate is important for all plants, especially when they are still young. Micro climates can be changed and improved by using good techniques, including all the techniques which have already been explained. For example rock swales, rocks will provide homes for small animals and insects, and at night, when the temperature is colder, the rocks will also become cold and moisture will collect on the rocks surface. This moisture will soak into the soil and be used by plants. This moisture is an important water source in dry areas.

8.6 STARTING REFORESTATION

Areas which are best for starting reforestation are areas which naturally have good micro climates. If you plant trees in these areas, the success rates of tree growth will be higher. Observe your land to know which areas naturally have good micro climates.

Look for:

⊙ Existing groups of trees. Trees will grow in a particular spot because the micro climate is better. Existing trees will provide mulch, shade and protection for newly planted trees

⊙ Grasses and small plants. In very dry areas, grasses and small plants indicate where the soil is better and where there is possibly more water available. Trees will grow better in these areas compared with other crops, because trees are more resistant. Areas with no grass indicate where the soil is very poor, with many rocks and not enough water

⊙ Groups of rocks. Trees planted below rocks will receive more water because the rocks will catch and direct rain water

⊙ Areas where water naturally collects

⊙ The northern side of a mountain. This is the best side of a mountain for reforestation because it receives the right amount of sunlight for trees to grow, and hence will have a better micro climate. But also observe which side is the most cleared or destroyed, and which side more urgently needs reforestation using night condensation

8.6.1 Assisting Natural Reforestation

Nature is always working towards a healthier environment. Don't work against nature, working with natural patterns will speed up the process. Some steps which work with nature towards shaping a healthier environment include:

- ◉ Stop burning. By burning, you are destroying many valuable resources. For example, burning grasses will also burn their functions, one of which is to protect newly planted trees

- ◉ Conserve bird habitats. Birds are very useful in reforestation, birds help to spread seeds through their manure. The manure will add nutrients to the soil and some of the seeds will grow into new trees

- ◉ First, plant trees in small groups. Then, in following years, add new trees to the existing groups. The new trees will receive protection and mulch from the older trees

8.6.2 Seed Balls

A seed ball is a small ball of clay, about 4 cm in diameter, containing plant seeds and dried manure. Seed balls are a good, simple technique to start reforestation in dry areas, steeply sloped areas, or areas with few or no plants or trees. Place the seed balls in any area you want before the wet season starts. The clay will protect the seeds inside from animals until the rains come. When the wet season starts the seeds will begin to grow and the dry manure will provide some nutrients to help them grow. It is best to use seeds of fast growing legumes, like acacia, leuceana and moringa. The trees that grow from the seed balls will improve the soil and provide protection and mulch for new trees planted afterwards.

Making Seed Balls

Choose clay that sticks together (doesn't break) when rolled into a snake shape. Add some water to the clay so that it becomes easy to shape into a ball. Mix in a small amount of manure, but make sure that the clay will still stick together. First, make the balls, then add about 5-10 seeds in each ball. The seeds must be inside the ball so that animals won't be able to eat them once they are in nature. Straight away, dry the balls in the sun for 1-2 hours. Leave until dry, but not cracked. Put them in a dry and shady place to continue drying. The balls must be completely dried because if they are still wet, the seeds will grow. When dry, store the balls in a dry place until you are ready to use them.

making seed balls

SMART IDEAS!

- Making small catchments of rocks for the seed balls will improve growth success rates because soil and water will collect there for the young trees

- Seed balls will help a lot if there is large areas of land you want to reforest, but have difficulty planting the whole land in one season. At the start of the wet season you can plant crops on the most productive part of the land, while seed balls can be used for other parts of the land making seed balls

8.6.4 Protection for the Reforestation Area

The reforestation area must be protected from fire, animals, strong winds and erosion. This protection will need community participation to work well. Neighbors and surrounding communities should be involved and should understand any reforestation project which affects them. Community group meetings can be held to discuss and plan together issues relating to protection for the reforestation area. Some community plans to develop together could be:

- Using traditional / community laws to increase the awareness of the whole community about the importance of reforestation and protection for the reforestation area

- Include schools, local groups, religious groups and government workers in the process of providing education for communities about the importance of reforestation and protection for the reforestation area

- Develop a sense of ownership in every community group member for shared community resources. These community resources include nurseries, cropland and community forests. This awareness is very important for increasing a communities ability to work together

- Develop short term and long term plans for protection of the reforestation land. Short term plans can be made for areas which need immediate attention or are more urgent Every idea and plan for community land management and protection for reforestation land should be discussed with the government. Working together with the government will improve results and increase community involvement.

SMART IDEAS!

Plan each activity well. In reforestation, it is better to work step by step, and make every small step a success, rather than trying to reforest a very large area of land, but not being able to manage it well.

8.6.5 Protection from Fire

Fire usually comes from the direction where the wind comes from in the dry season or from areas lower down the mountain. Make fire protection on your land in these area. Fire protection could be:

- Living fences made from plants or trees which are fire resistant, such as cactus, aloe vera and banana

- Rock walls. Besides functioning as fire protection these walls will also act as a wall to stop animals from entering

- Firebreaks. Firebreaks is a bare strip of land which is kept clear of plants. When a fire reaches this area, it will go out because there is nothing to burn These techniques will all work better if they are combined.

8.6.6 Protection from Animals

Animals like goats, buffalo, cows and pigs can damage large numbers of trees very quickly. To avoid this, make small fences or tree guards surrounding each tree. Fences can be made from any inexpensive and available materials, such as wood, bamboo, rock, wire, net, or a combination of materials. Living fences made of plants which animals don't like, such as cactus, will also provide protection from animals.

Tree guards are good to use for fruit trees, house trees and large trees which are still young. Once trees grow tall enough and their leaves are above animal reach, the tree guards can be removed and animals can be left free in this area.

8.6.7 Protection from Strong Winds

If plants are protected from strong winds, they will grow faster and healthier, especially when they are still young. Protection from strong winds could be living fences, vine trellises or trees planted to form a windbreak.

For croplands, plant a few lines of trees specifically to function as a windbreak. These trees can be of many different types, from legumes to fruit trees. Plant the line of trees in the direction where strong winds most often come from. For reforestation land, first plant groups of trees in areas which are already protected from strong winds. In years following, add new trees to the existing group. The new trees will be protected by the established trees.

8.6.8 Protection from Erosion

Planting trees is the best long term solution to prevent erosion, but when the trees are still young, they will also need protection from erosion. This protection can include many techniques which have already been explained, such as using swales, terraces and more. Grasses, bush and ground cover crops will also help to prevent erosion.

8.6.8.1 Planting Trees

The techniques used to plant a tree are very important for the tree growth later. Some techniques to use are as simple as:

- Planting in the afternoon. Don't plant trees during the heat of the day
- Supplying enough water
- Take care of the roots as much as possible, don't disturb them
- Make a small trench surrounding newly planted trees for catching water. A watering pipe can also be added and will work even better
- Put mulch around the tree

8.6.8.2 Planting Fruit Trees

If you have enough water, fruit trees can be planted at any time of the year. If water is limited, it is best to plant when the soil is wet or at the start of the wet season.

8.6.8.3 Techniques for planting fruit trees:

⦿ Dig a hole knee deep, or more if possible. Fill the hole with water. Also water the tree when still in its container

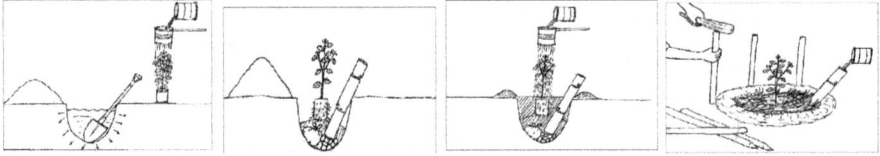

⦿ Put a pipe (which can be made from bamboo) inside the hole. Place some gravel below the pipe to help with water flow later on

⦿ Fill a plastic bag with manure and place it at the bottom of the hole. If available, use a bag made of natural materials which will still hold the manure for a long time in the soil. Cover with soil and make a small mound in the hole for the tree to sit on

⦿ Carefully, remove the tree from its container without breaking its roots. If there are many roots, gently loosen the bottom tree roots. Then, place the tree in the hole which has been prepared

⦿ Fill the hole with soil. Make a shallow trench around the surface for water collection and to help with water supply. Make sure that the top of the tree roots are covered with at least 2 cm of soil to prevent the roots from drying out

⦿ Add lots of mulch around the tree

⦿ Water enough

⦿ Make tree guards if needed

8.6.8.4 Planting Reforestation Trees

Reforestation trees are planted using the almost the same techniques as used to plant fruit trees, with a few small changes.

This is because:

⦿ Reforestation trees are usually planted further away from the garden and house area

⦿ Reforestation trees need less fertilizer

⦿ Reforestation trees are not usually watered so rain water storage is very important

⦿ The ground is often harder, making it more difficult to dig

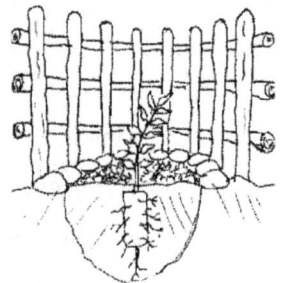

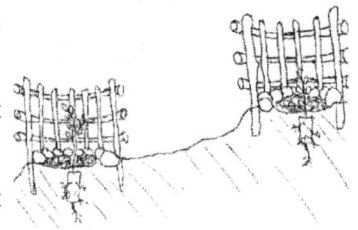

The best time to plant reforestation trees is at the start of the wet season, when it begins to rain consistently.

Follow the same steps as used for fruit trees, but with these few small changes:

⊙ Dig a smaller hole

⊙ There is no need for a bag of manure in the hole

⊙ Make a large trench for water catchment. Make sure that the trench is above ground level. This will help to prevent too much water collecting during the wet season

⊙ Use watering pipes during the dry season

SMART IDEAS!

⊙ Dig holes for the trees, but leave them empty until the rains come. The rain water will collect in these holes and soften the soil so when trees are planted, they will grow better

⊙ Planting with swales will always improve results and help trees to grow faster and healthier

8.7 Tree Maintenance

8.7.1 Watering

Fruit Trees and Tree Crops

Fruit trees and tree crops must be regularly watered during the dry season to achieve good production and larger fruits, especially in the first few years. Here are some suggestions for watering:

⊙ It is better to water trees with a lot of water very week than with a small amount of water every day or two. This will encourage roots to grow down farther looking for water so they will reach ground water faster

- ◉ Use watering pipes
- ◉ Water trees in morning or afternoon

8.7.2 Fertilizer

Plants use nutrients in the soil to live. Therefore, nutrients in the soil which are used by plants need to be replaced so that the plants will grow healthy and produce the best it can. The same is true for people and animals, but luckily trees don't have to eat every day.

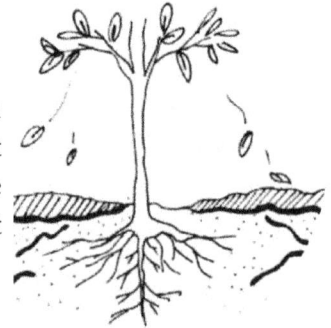

Fruit Trees and Tree Crops

Compost, liquid compost, manure and mulch provide many different nutrients and other benefits as well. The best place to fertilize trees is where the roots soak up nutrients. Underneath the outside leaves of every tree is the 'root feeding zone'. This is where the plants outside roots are and where the tree will most easily be able to use nutrients. A small circular mound surrounding the root feeding zone will improve watering and fertilizing results. This mound circle can be enlarged as the tree grows.

Watering pipes can also be used to feed liquid compost directly to the trees roots in the ground.

Fertilizers which work best for fruit trees and tree crops are:

- ◉ **Compost and Manure.** Compost and manure can be applied twice a year, just before the wet season starts and at the end of the wet season. Apply to the root feeding zone. Use about a 5 cm layer (the length of one finger) of compost or mulch, especially around the root feeding zone. This will provide many important nutrients for the tree

- ◉ **Liquid Compost.** For trees up to 3 years old, use about 1 gallon (20 liters) of liquid compost, for trees over 3 years old, use about 3 gallons. Put some of it through watering pipes and some directly on the ground over the root feeding zone. Use once every 2 months during the wet season and only once in the middle of the dry season

- ◉ **Mulch.** Apply mulch just outside the root feeding zone, closer to the tree trunk. Don't let mulch touch the tree trunk, because if it does, disease or fungus could damage the tree. Leave about 10 cm of space from the tree trunk. Use a thick layer of mulch to keep the ground moist and to improve the soil quality more quickly. If available, seaweed makes a very good mulch for trees, but wash it first to remove excess salt

- ◉ **Urine.** Urine is also a good source of nutrients because it contains lots of nitrogen and is constantly available. Citrus trees especially like urine fertilizer. Before applying on plants, urine should be diluted in a bucket of water. This can be applied more often for established trees, but not too often for young trees

Reforestation Trees

Reforestation trees need less fertilizer than fruit trees, and fertilizing is most important when the tree is still young. On reforestation lands, the available nutrients are often not enough for plant growth. Good natural fertilizing techniques will replace these lost nutrients quickly.

Natural fertilizing techniques which can be used include:

- ⊙ Compost, manure or seaweed. These can be applied when planting to provide some nutrients for the young trees

- ⊙ Legume trees are an important source of nutrients. Their roots provide nutrients and the trees can be pruned up to 5 times during the wet season, which will provide mulch materials. These trees can also be used as 'pioneer trees' and as mulch and nitrogen providers for other plants

- ⊙ Mulch. For reforestation trees, mulch provides many nutrients which trees need

- ⊙ After 3 years, animals can be carefully introduced to the reforestation land.

8.7.3 Mulching Trees

Mulching is an important part of tree maintenance.

Mulch provides many benefits, including:

- ⊙ Holds water in the ground and helps to keep the ground moist for longer

- ⊙ Maximizes the benefits of manure and compost if mulch is used as a top layer

- ⊙ Acts as an important source of nutrients for trees

- ⊙ Improves soil quality by increasing organic matter and soil biota in the soil

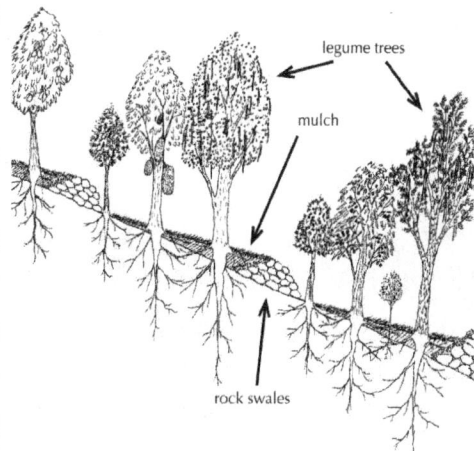

Fruit Trees and Tree Crops

Continuously apply mulch to trees. A layer of up to 10 cm or more will give the best results. To avoid fungus or disease, don't let mulch touch the tree trunk. Organic materials which can be used as mulch include rice husks, coffee husks, seaweed (it must be washed first), tree cuttings, dry grass and weeds, and even used paper, boxes, wood and bamboo will make good mulch.

Compost and dried manure will provide more benefits to soil and plants if they are placed under a layer of mulch.

Reforestation Trees

Natural mulch, such as leaves, grasses and weeds, will quickly form into mulch if the land is not burned. This mulch can be collected and placed around trees. Legume trees and other trees can also be pruned to provide more mulch materials.

Mulch will provide the most benefits if combined with swale systems. Rocks can also be used as a mulch, especially for dry areas because they will provide extra water for your trees during the dry season. For more information, see the micro climate section in this module.

BEWARE!

During the dry season, mulch will burn easily and can cause fires to spread. Therefore, burning the land must be stopped and fire should be prevented from entering your land. Community participation and understanding, as well as using practical techniques, is very important.

8.7.4 Tree Pruning

Pruning trees is important for maintaining the tree's health and productivity. Disease and fungus can spread easily if trees are not pruned. By pruning, harvesting will also be much easier because the tree will be lower to the ground and easier to access.

Using the right technique to prune trees is very important so that the tree will grow back quickly, not suffer from stress and be able to avoid disease and fungus.

When pruning tree branches use a saw or very sharp machete. Prune the branches as close to the trunk or main branch as possible. Make the cut as smooth as possible, angled and not flat, because a flat cut will increase chances of disease.

For fruit trees and other tree crops, you can paint a layer of jackfruit sap on the cut. This will stop disease or fungus from entering through the cut.

Fruit Trees and Tree Crops

For almost all tropical fruit trees, it is recommended to only prune when necessary.

Some reasons for pruning are:

- ◉ To remove dead or diseased branches. By removing dead or diseased branches, you will reduce the chances of fungus and disease spreading. Cut the branches off before the diseased or dead part begins. The pruned branches should be taken away or burned to avoid spreading more disease

- To allow more sunlight into the middle of the trees. This will increase fruit production and reduce fungus problems.

- Only prune if necessary and don't prune the tips of all the branches, but only some of the longer branches

- To encourage new growth on older trees. When trees have grown old and are not producing well, pruning them will encourage new growth and better fruit production

Reforestation Trees

Reforestation trees need less maintenance and pruning than fruit trees, however some pruning will help improve growth and quality.

Some reasons for pruning reforestation trees are:

- To get firewood. Dead or diseased branches can be removed and used as firewood

- Removing lower branches will make more room for people to walk around, for animals to graze and to grow other crops underneath, like coffee and vanilla. Don't remove lower branches of windbreak trees because this will make them less effective

8.8 LEGUMES

Legume trees which are planted for mulch, soil improvement or as living fences should be pruned back regularly. By pruning legume trees, you will be returning nitrogen back to the soil through the tree's roots. Prune back the trees until they are about waist height to make maintenance easier, prune the whole tree evenly for best results.

8.9 BAMBOO

- Bamboo grows in all areas and has many important uses for communities. Generally, bamboo can be split into 2 categories: Clump bamboo (sympodial) and creeping bamboo (monopodial).

- Clump bamboo grows in tropical climates and is more common in our climate, while creeping bamboo generally grows in subtropical climates.

8.9.1 Bamboo provides:

- ⦿ Income
- ⦿ Building materials
- ⦿ Furniture materials
- ⦿ Food for people and animals
- ⦿ Fences, living fences or trellises
- ⦿ Windbreaks
- ⦿ Irrigation pipes
- ⦿ Bamboo charcoal for cooking
- ⦿ Material for making musical instruments
- ⦿ Material for making containers
- ⦿ Material for handicrafts and much more

The process of planting and managing bamboo clumps properly is the first step for producing high quality and easy to harvest bamboo.

8.9.2 Bamboo Propagation

There are a few techniques for bamboo propagation, including rhizome propagation, using branch cuttings or using branches and seedlings for some larger types of bamboo.

The technique you will use depends on what type of bamboo you are propagating and what the bamboo will be used for.

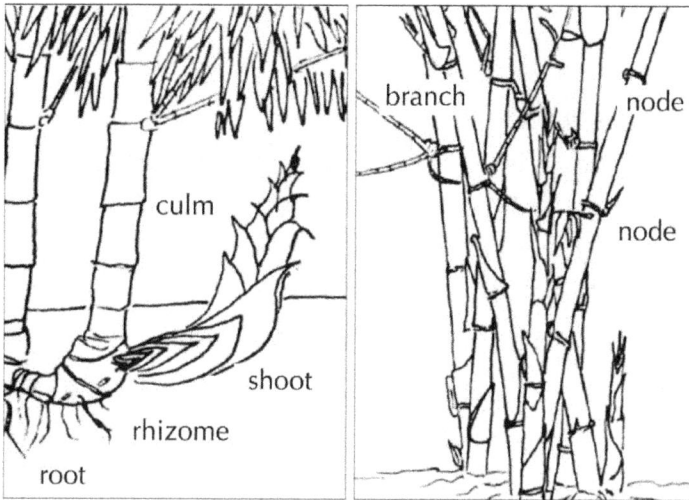

For drier areas, the start of the wet season is the best time to propagate bamboo. However, if enough water is available, propagation can be done at any time.

8.9.3 Rhizome Propagation

Rhizome propagation is good for small scale planting because it has a high success rate. However, this technique is more difficult and takes more time than other techniques. Rhizome propagation works with almost all types of bamboo, but rhizomes of large species are usually too difficult to dig up. Therefore, rhizome propagation works best with smaller bamboo species with many rhizomes and culms.

Rhizome propagations steps:

- ◉ Choose the bamboo rhizome and culm which you want to propagate, one year old culms on the outside of the clump are easiest and best to use

- ◉ Cut the culm three or four nodes above ground level 3. Cut again through the rhizome, where the rhizome joins with the next rhizome. Usually this is towards the center of the clump. Dig the roots and soil about 10-15 cm away from the culm, so that when you remove the rhizome, the roots and soil stay attached

- ◉ Keep the rhizome and roots wet until planting or plant immediately. Wet the leaves as well. Keep the rhizome and roots out of sunlight

- ◉ Plant the rhizome about 15 cm in the ground and water well. Apply fertilizer and compost, then add a layer of mulch around it New leaves and branches will grow from the bamboo and at the beginning of the wet season, new shoots will grow from the rhizome. Sometimes new shoots will grow right away.

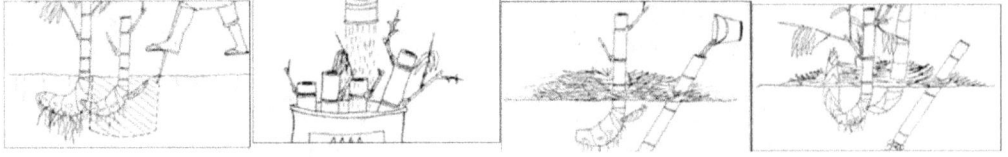

Culm (Pole) Cutting Propagation

Culm cutting propagation is better for large scale planting and windbreaks because it is easier and takes less time. However, the success rate is lower. This technique is best to use with larger bamboos, which are difficult to propagate using rhizomes.

Culm cutting propagation steps:

- Choose a bamboo culm with lots of branches, aged 2-3 years

- Cut it as close as possible to the ground, then cut the culm into 1.5-2 meter lengths

- Cut off the branches and leaves after the first node on each branch, leaving only 2 or 3 branches on one side

- Dig trenches and bury the bamboo culms about 15 cm in the ground. After planting, cut the remaining branches at 2 nodes above ground. This will help you to see where the bamboos are planted

- Water every day for the first week. After that, water twice a week for one month When the culms begin to grow new shoots, they can be dug up, cut and replanted wherever you want them.

Branch Propagation

Choose a few larger bamboo branches, they are usually at the top of grown bamboo. Cut the branches as close as possible to the main branch, about 1 meter long (there should be a minimum of 3 nodes on each branch). Plant the branches in healthy soil, and treat the same as other plant cuttings. It is best if planted at an angle.

8.9.3 Bamboo Nurseries

Culm and branch propagation can also be done in containers. But rhizomes don't grow well in containers and should be planted straight into the ground.

SMART IDEAS!

- Don't use the top 1/3 of each culm, because the success rates will be much lower

- Cut a hole between each node before burying or planting to help hold water

Growing High Quality Bamboo

Every type of bamboo has a different quality and character. Growing different species of bamboo in one area will give many benefits to people because they will receive different benefits to fit their different needs. Each person in a community can plant one type of bamboo, and then they can exchange the different types of bamboo.

To grow high quality bamboo, it is important to supply the plants with enough nutrients, and remember that bamboo plants are heavy feeders. Bamboo has root systems which grow close to the grounds surface. Because of this, it is best to give bamboo small amounts of fertilizer regularly, for example every 3-6 months rather than large amounts of fertilizer once a year.

exchanging bamboo seedlings

The best fertilizers to use are manure and compost, especially just before the wet season. The best manure to use is pig manure, it contains all the nutrients needed for bamboo growth. Applying a layer of mulch, about 30 cm thick, will also improve bamboo growth. When the plants are 2 years old, thinly sprinkle cement powder around the clump (underneath the mulch).

Cement contains silica, a mineral which will help the bamboo harden and improve bamboo pole quality.

Experiments done with bamboo show that bamboo timber is stronger if grown on hillsides rather than near rivers.

Clump Management

A properly managed bamboo clump will produce high quality bamboo and will be easy to harvest. A well managed clump of bamboo will have a range of different aged culms, from 3 years, 2 years, 1 year and new shoots. There should be 6-8 culms of each age in every bamboo clump, which makes 24-

an unmanaged bamboo clump

32 culms of bamboo per clump. They should all have enough space to grow well and be easy enough to harvest.

Opening Clumps

A well managed clump of bamboo will look open and healthy, which makes it easier for us to choose and observe which bamboo is ready for harvesting and which bamboo is still too young. An unmanaged clump of bamboo will look tightly packed and disorganized, making it difficult to choose which bamboo is ready for harvest, so there will often be dead or dry culms in the middle of the clump. This type of condition will make it difficult for us to harvest.

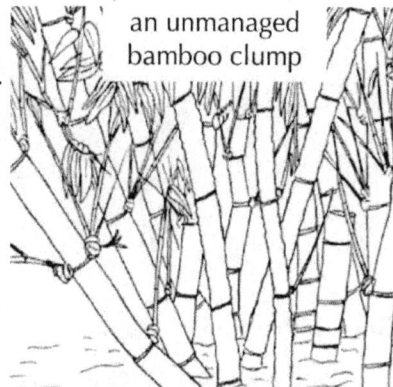

The first step in managing a bamboo clump is by cutting or removing all the old or dead culms. This will be difficult because sometimes they are located in the middle of the clump. One way to do this is to cut into one side of the clump till the middle, then cut out all the old or dead culms. Cut them as close as possible to the ground. This will create a shape that allows us to harvest mature shoots from the center of the clump without damaging new shoots which are usually located outside of the clump.

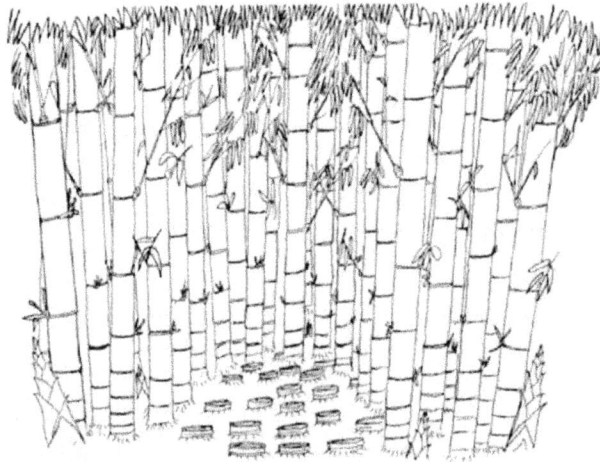

Thinning

Remove any damaged or bent culms and any culms which are growing too close together. If the clump has been harvested before, there will be many culm stumps, these stumps should be removed, cut them as close as possible to the ground. This will make it easier for us to reach the middle of the clump.

Branch Pruning

Prune off lower branches to make clump access easier. Cut the branches at the second or third node to avoid fungus reaching the culm.

Choosing and Marking New Shoots

During shoot season, choose 6-8 healthy shoots, located in a good position. Remove all the other shoots, this will encourage new shoot growth in the future. The removed shoots can be used as vegetables or animal feed.

The chosen shoots can be marked to keep track of their age. The bamboo poles will be stronger harder and more insect resistant if they are harvested at 3 years older or more. Mark the shoots by scratching a number into the shoots before their leaves grow, this scratch will leave a permanent mark. Mark all the shoots when they are about the same height, about 1 meter above ground level is good. For example, for the year 2004, mark all shoots with the number 4, the bamboo will be ready to harvest in the year 2007, so you will know that all shoots with the number 4 are 3 years old.

8.10 BAMBOO PLANTATIONS

Bamboo can be grown near the house, on cropland or as part of a managed system. A bamboo plantation is the most efficient way to produce high quality bamboo. Produce from a bamboo plantation will fulfill many functions, it will provide shoots for vegetables, leaves for animal fodder and bamboo for charcoal, and the bamboo clumps can also function as windbreaks, living fences and provide erosion control.

8.10.1 Intensive Plantations

- Intensive bamboo plantations are plantations where bamboo is the main crop. The bamboo can be planted in rows, with 4-6 meters between clumps and 8-10 meters between rows. On sloped lands, the bamboo should be planted on contour.

- By leaving 8-10 meters of space between rows, there will be enough room to harvest and collect the poles. You can also graze animals between these rows.

8.10.2 Mixed Plantations

Mixed plantations are plantations where bamboo is one of many different types of crops, for example a coffee plantation with bamboo functioning as living fences and windbreaks. Bamboo can be combined with crops of about the same height, like mango, coconut, avocado, jackfruit, timber trees and fibre trees. The combination of plants used is up to you, but don't forget to leave enough space for harvesting bamboo in the future. Animals can also be integrated into this system. Planting bamboo on hillside contours will help prevent erosion and stabilize the edge of a terrace.

8.10.3 High Quality Bamboo Poles

Producing high quality bamboo poles will depend on the following factors:

- ◉ Bamboo species
- ◉ Bamboo pole age
- ◉ Harvesting time
- ◉ Curing and storage
- ◉ Preservation

1. Bamboo Species

Some types of bamboo are naturally stronger and more resistant to borer insects than other types of bamboo. In Indonesia, the types of bamboo which are commonly grown and used include betung / petung bamboo, tali / apus bamboo, gombong bamboo, item bamboo, ampel bamboo, duri bamboo, santong bamboo, tutul bamboo, yellow bamboo and more.

2. Bamboo Pole Age

Bamboo poles should be harvested when they are at least 3 years old. For some species of bamboo, it is better to harvest at 4, 5 or even 6 years old. Tali / apus bamboo is best to harvest after 3 years, but petung bamboo should only be harvested after 4 or 5 years.

If bamboo poles are still 1-2 years old they contain more compound sugar / starch, which borers and starch insects (Dinoderus sp) like to feed on. After 3 years there is less starch and silica becomes more dominant. Silica is a mineral which makes bamboo poles harder and more resistant to insects. Bamboo harvested under 3 years will shrink and crack easier, and attract more borers and starch insects. Bamboo harvested after 3 years will be stronger and more insect resistant.

3. Harvesting Time

The best time to harvest bamboo is during the dry season. Choose a time when new shoots are almost at their maximum height and have just begun to grow leaves at the top. At this time mature bamboo will be in its strongest condition. A common practice in Asia is to harvest bamboo on the full moon. This is to help prevent borers in the bamboo, and the bamboo will contain less moisture during the full moon. Following this practice will produce better quality bamboo. Avoid harvesting during shoot season, because at this time the bamboo are still 'feeding' their young. At this time the bamboo will contain high amounts of water and sugar. And besides that, cutting bamboo at this time will damage the new shoots.

8.10.4 Curing and Storage

Bamboo needs 4-8 weeks to dry before it is used. If bamboo is stored vertical it will take about 4 weeks, while if it is stored horizontal it will take about 8 weeks. Bamboo must be cured and stored in the shade, not touching the ground and out of the rain.

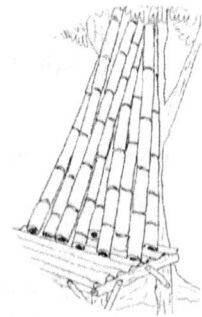

8.10.5 Preservation

Borers, fungus and termites are the biggest problem with bamboo. You need to preserve bamboo to make it more resistant to these insect pests. It is also important to understand how borers work. Borers are small beetles which lay their eggs in damaged parts of bamboo skin. This could be at the ends where it was cut, where branches have been removed or where the skin has been scratched. The borer eggs will hatch at different times and the borers will then eat the compound sugar / starch inside the bamboo. Therefore, borer attacks can be prevented with good management and by not damaging the bamboo poles.

The first step in preserving bamboo is to reduce the amount of starch in the bamboo. This is why you should only harvest bamboo during the dry season and only after the bamboo is aged 3 years or more. The amount of starch in bamboo is lowest during the dry season and in older culms.

The next step is to reduce the starch content even further. This can be done in many differentways, including:

- **Clump drying.** The poles can be cut and left in the clump for 4-6 weeks, until their leaves have all fallen. The pole should be placed on rocks so they are not touching the ground. The leaves will use up most of the starch in the pole and the pole will dry slowly without any areas for borers to lay their eggs

- **Preserving with water.** The poles can be soaked in running water for 2-3 weeks. The water will clean out most of the starch. After soaking in water, the bamboo poles must be dried slowly in the shade. Don't dry in the sun because the bamboo poles will crack

- **Preserving with seawater.** For treatment with seawater, the bamboo poles can be soaked directly in the ocean. Tie the bamboo tightly to rocks so they won't float away with the tides. Don't let the bamboo poles lay exposed to sun at low tides, because they will crack Another method, which is perhaps easier, is to dig a pit on land near the ocean. The pit will naturally fill with seawater as you dig below sea level.

With both of these methods, leave the bamboo soaking for 2 weeks. Afterwards, remove the poles and leave them to dry in the shade.

- Tuha treatment. There is a type of plant called "tuha" which can be used to preserve bamboo. Tuha is poisonous for people and animals, so it must be used carefully. To use tuha, make a solution of 1 bucket of tuha combined with 200 liters of water.

- You can use an old drum as a container. Short pieces of bamboo can be cured in the drum, and poles of bamboo which have just been harvested can be placed in the drum with their leaves still attached. The liquid in the drum will be drawn up through the pole to the leaves. Add more tuha into the drum as needed, then leave for one week. After this time, remove the leaves, and take the poles out of the drum to dry in a shady place off the ground. You can also use a tank or container made specifically for curing bamboo. This container should have a cover to stop rain

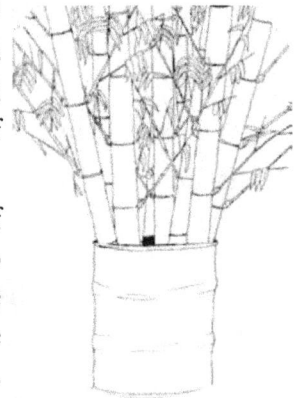

from seeping in and children or animals from entering. Cut the bamboo in lengths and remove the branches from the poles. Put the poles in the container filled with tuha liquid and leave for 4-6 weeks. Then, remove the bamboo poles and dry them in a shady place until they will be used. If you use a water treatment first, than the tuha treatment will only take 2 weeks.

- ◉ Oil and varnish. Finished pieces of furniture or crafts can be oiled or varnished to prevent fungus from growing, make the product last longer and increase the value of the product

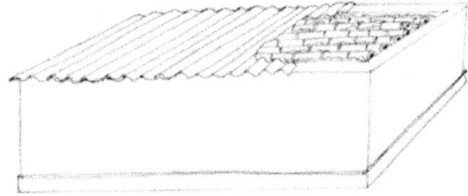

- ◉ Preserving with Borax. Bamboo can also be treated with borax, a chemical which will kill borers and their eggs. Borax is best to use on large amounts of bamboo which need to be cured quickly, this is usually for export purposes. Most countries will not import bamboo which has not been chemically treated. The way to treat bamboo with borax, is to simply soak bamboo poles in a borax solution for 2 weeks, then dry them in a shady place off the ground

8.10.7 Precautions

- ◉ Borax is a very strong chemical. When using borax, you should wear protective clothing, and afterwards always wash thoroughly

- ◉ Borax solutions should be disposed of carefully. If the borax is diluted in water, the solution can be spread around fruit trees. Spread as widely as possible. This solution contains mild pesticide and herbicide properties. If diluted to 1% (1 part borax diluted in 100 parts water), it can be used on vegetable gardens. Do not dispose of this solution in rivers or irrigation systems

8.11 USING BAMBOO

8.11.1 Building Materials

Bamboo can be used for building houses, walls, floors, roofs, animal pens and much more. Bamboo is a strong, lightweight and easy to use material. Bamboo is also very decorative and can be used to make the house more beautiful.

8.11.2 Furniture Materials

Bamboo furniture is very beautiful and long lasting, especially if the bamboo used has been treated properly. Bamboo can be used for making chairs, tables, beds, wall panels, shelves and much more. To learn and build furniture requires training, tools and imagination.

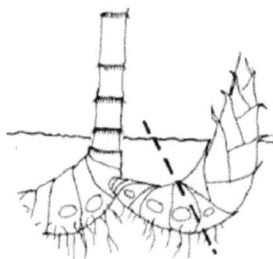

8.11.3 Food

Food for People

Bamboo is highly nutritious food, it contains water, carbohydrates, amino acids and many vitamins and minerals, and it can be cooked in many different ways. Bamboo is commonly eaten in many Asian countries. Some bamboo produces edible shoots, and other types produce shoots which are not good for eating. Some types of bamboo shoots which can be eaten include petung / betung bamboo, hitam bamboo and tabah / tawar bamboo. In the wet season, new bamboo shoots will grow in bamboo clumps. The new shoots are the edible part of bamboo. Cut the new shoots near the bottom where they become hard. The harder parts bamboo shoots will taste bitter. The best part of shoots to eat is the inner parts, which are usually white in color. This part is soft, tastes good, and will be easy to cook with many dishes. Bamboo shoots can also be pickled, dried or fermented to make them last longer.

> *Don't harvest bamboo shoots for eating until the bamboo plant is over 3 years old. Before this time, cutting the shoots will damage the root system and cause the bamboo to grow much slower.*

Food for Animals

Bamboo shoots are also good food for animals, especially for pigs. Cook the shoots together with other materials, like cassava, sweet potato, leaves and so on. Bamboo leaves and stalks are also quality animal food, especially for goats and cows, which will benefit from the silica content in the bamboo leaves and stalks.

8.11.4 Fences

Bamboo is a common material used for fences, both as living fences and as fencing material. If used as fencing material, bamboo should be used for the crosspiece, not for the posts which are in the ground, because if bamboo is in the ground it will rot much easier.

Living Fences

Bamboo plants will function well as living fences. It will take a few years for the bamboo the become thick enough, so temporary fencing will need to be built beside the bamboo plants. Bamboo living fences are good to use for animal yards, including for chickens, ducks, cows, buffaloes and pigs. The bamboo will provide shade and food, and can function as a windbreak. Bamboo can also be grown around orchards, but should not be grown to close to vegetable gardens because the bamboo will soak up lots of water and nutrients and may give too much shade.

8.11.5 Trellising

Bamboo can easily be shaped into just about anything, including trellising, because it is light and easy to move. The trellis can be made in any shape to fit your needs. Bamboo is also decorative, it can add more beauty to your garden.

8.11.6 Windbreaks

If bamboo is planted close together, it will eventually form a fence. Bamboo clumps can also function as windbreaks.

8.11.7 Irrigation Pipes

There are many ways to use bamboo for irrigation:

- ◉ Bamboo which has been split in half with nodes removed is commonly used for flowing and directing water

- ◉ Bamboo poles can be cut into 1 meter lengths and placed in the ground for watering fruit trees and

- ◉ Vegetables. Put holes in the bamboo's inner nodes to allow water through. This technique will save a lot of water and improve plant growth

⦿ Bamboo can be used as pipes, which are useful for many purposes

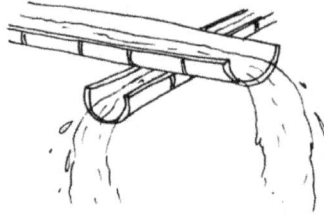

(For more information about irrigation, see Module 6 – Home & Community Gardens).

8.11.8 Bamboo Charcoal for Cooking

Bamboo charcoal can be made and used for cooking as a substitute for firewood. The charcoal is made from pieces of burnt bamboo, arrowroot powder and water. Bamboo charcoal will produce heat well, without producing a lot of smoke. Using bamboo charcoal is much easier than collecting firewood. (For more information about bamboo charcoal, see Module 12 – Appropriate Technology).

8.11.9 Musical Instruments

Bamboo is a good material for making many different musical instruments, such as flutes, wind chimes and shakers.

8.11.10 Containers

Cooking Containers

Bamboo is traditionally used as containers for cooking meats and vegetables.

Bamboo Buckets

Large bamboo poles can be used to make buckets and watering containers. Bamboo buckets or containers will last much longer if they are varnished before use.

Storage Containers

Bamboo can be easily used to make containers for anything, such as containers for jewelry, writing materials, cooking utensils, flower pots and even seed storage containers.

These containers can be easily decorated, carved or shaped, and can then be marketed for sale. The containers will also last a long time if they are treated properly.

If the containers are used for seeds, the bamboo should be treated to prevent pest problems. However, don't use bamboo which has been treated with tuha for storing food, because tuha is poisonous.

Pots / Plant Containers

Small bamboo poles can be cut into pieces and used as seedling containers. Larger bamboo poles can be used as pots for flowers, spices and house plants. Don't forget to put a few small holes at the pots base to allow water drainage.

is poisonous.

REFERENCES

Hemenway, Toby. Gaia's Garden: A Guide to Home--Scale Permaculture. Second Edition. Chelsea Green Publishing Company, 2009.

Holmgren, David. Permaculture: Principles & Pathways Beyond Sustainability. Holmgren Design Services, 2002.

Holzer, Sepp. Sepp Holzer's Permaculture: A Practical Guide to Small--Scale, Integrative Farming and Gardening. Chelsea Green Publishing Company, 2011.

Mollison, Bill and Slay, Reny Mia. Introduction to Permaculture. Tagari Publications, 1997.

Mollison, Bill. Permaculture One: A Perennial Agriculture for Human Settlements. Tagari Publications, 1981.

Mollison, Bill. Permaculture Two: Practical Design for Town and Country in Permanent Agriculture. Tagari Publications, 1979.

Mollison, Bill. Permaculture: A Designers' Manual. Tagari Publications, 1988.

Perkins, Richard. Why Permaculture Needs Accurate Data and Measurement to Persuade the Mainstream. Permaculture – Inspiration for Sustainable Living. May 2nd 2012.

Permaculture Activist Magazine: www.permacultureactivist.net.

Pimentel, D., S. Williamson, C. Alexander, O. Gonzalez--Pagan, C. Kontak, and S. Mulkey. 2008.

Reducing Energy Inputs in the US Food System. Human Ecology 36 (2008): 459--71.

Senge, Peter. 2006. The Fifth Discipline. Doubleday, NY, p 12.

Shepard, Mark. Restoration Agriculture: Redesigning Agriculture in Nature's Image. Acres U.S.A., 2013.

Weiseman, Halsey and Ruddock. Integrated Forest Gardening: The Complete Guide to Polycultures and Plant Guilds in Permaculture Systems. Chelsea Green Publishing, 2014.

INDEX

www.ingramcontent.com/pod-product-compliance
Lightning Source LLC
Chambersburg PA
CBHW082006190326

41458CB00010B/3089